AGENTIC AI + ZERO TRUST

A Guide for Business Leaders

JOSH WOODRUFF

MICHELLE SAVAGE

ISBN: 979-8-9995300-1-1 (Paperback)
ISBN: 979-8-9995300-2-8 (Hardcover)
ISBN: 979-8-9995300-0-4 (eBook)

DISCLAIMER: The strategies and approaches in this book reflect the author's experience and expertise. For specific implementation in your organization, consider engaging qualified professionals.

First Edition: September 2025

Published by:
MassiveScale.AI
AI & Security Advisory
New York, NY
massivescale.ai

Printed and distributed worldwide

10 9 8 7 6 5 4 3 2 1

Trademark Notice Company names and product names mentioned herein are trademarks of their respective owners.

Bulk Sales Information For information about bulk purchases, please contact sales@massivescale.ai.

CONTENTS

Foreword

By John Kindervag, Creator of Zero Trust

I've always enjoyed my conversations with Josh, so when he asked me to write this foreword, I enthusiastically accepted. Not just because of what this book represents for AI security, but because it shows how to keep humans in control when digital workers can make millions of decisions without us. That's the real challenge—and opportunity—of agentic AI.

You see, I'm at a point in my career where I'm thinking less about creating new frameworks and more about who will carry them forward. I'm relying on people like Josh and others like Chase Cunningham and George Finney to take these principles and apply them to challenges I never imagined when I first articulated "never trust, always verify" fifteen years ago. My goal these days is simple: ensure Zero Trust continues to evolve in the right hands while I play with my grandson.

How do you turn over a mantle? By finding people who understand the principles deeply enough to apply them to entirely new problems—and who can explain them in ways everyone can understand. That's exactly what Josh has done with this book.

Josh told me his wife Michelle wouldn't let him use any jargon—she'd "beat him up" if he tried. That's exactly what this book needed. Most security books are written by security people for security people, and they're inaccessible if you're not already an expert. This book is different. You can actually understand it whether you're a CEO, a board member, or just someone trying to figure out what AI means for your business. It's a conversation, not a lecture.

When I created Zero Trust, we were worried about human users with stolen credentials. Now Josh is applying those same principles to AI agents that can make millions of decisions per second. The strategy hasn't changed—the implementation has evolved dramatically. That's how you know a framework has lasting value: when others can take it places you never imagined.

Here's what I appreciate about Josh's approach: he's not trying to reinvent Zero Trust or add unnecessary complexity. He's taking the fundamental question—"What am I trying to protect?"—and applying it to a world where the threats learn and evolve autonomously. He understands that all bad things happen inside of *allow rules,* whether those rules govern humans or AI agents. The difference is that when an AI agent exploits an allow rule, it can do so millions of times before you notice. That's why the principles in this book matter so much.

Nobody really knows what AI is yet. We're at the cooing stage of AI, not even the talking stage. We're all making educated guesses about where this technology will take us. But Josh asks the right questions: How do we maintain human agency over systems that can think faster than us? How do we set boundaries for entities that find creative ways around them? How do we verify trust for agents that evolve every day?

But here's what I do know: the whole point of AI has to be about making life better for humans. Because that's what life is—life is humans. These digital systems we're building have no value unless they serve human purposes. If a machine has agency over you, that's a bad thing. Always. 100% of the time. You need agency over the AI, not the other way around.

This isn't a book with all the answers. Good books raise more questions than they answer. They start conversations between the author and reader, sparking connections that lead to new insights. As I've said, no book has all the answers—the best books have great chapters that make you think, that send you looking up other things, connecting dots you hadn't seen before. Josh has written that kind of book—one that asks the questions that need asking about agentic AI and Zero Trust.

The machines aren't taking over. But they are taking on more responsibility every day. AI agents are already out there making decisions, processing transactions, diagnosing problems. The question isn't whether to use them—competitive pressure will force that decision. The question is whether you'll maintain agency over them or cede it through poor design and wishful thinking. You must be able to tell these systems what to do, set their boundaries, and maintain control. They work for you, not the other way around.

For business leaders wondering whether to trust autonomous agents with critical decisions, this book provides a framework for doing so safely. For security professionals trying to protect systems that learn and evolve, it offers practical patterns that work. For anyone trying to make sense of AI's role in their organization, it provides clarity without false certainty.

What excites me most is that Josh and others aren't just maintaining Zero Trust—they're extending it into territories that demand new thinking. They're asking questions I wouldn't have known to ask. They're solving problems that didn't exist when I was focused on traditional perimeter security.

This book represents exactly what I hoped would happen: thoughtful people taking foundational principles and applying them to make life better for human beings in new ways. Because that's what all of this is about—making life better for human beings. If that isn't your goal, then your goal is completely screwed up.

Most importantly, it shows that Zero Trust isn't just about the past—it's about the future. A future where Josh and others carry these principles forward, applying them to challenges we can't yet imagine, ensuring that no matter how sophisticated our tools become, they remain under human control and serve human purposes.

I can retire happy knowing that people like Josh won't screw it up. In fact, they're making it better than I ever could have imagined.

Read this book. Apply these principles. Carry the torch forward. The future of secure AI—and Zero Trust itself—is in good hands.

John Kindervag
Creator of Zero Trust, Chief Evangelist at Illumio

Preface

This book began with a simple interview. Michelle, my wife who also happens to be a writer, was interviewing me for what was supposed to be a quick blog post about agentic AI—just as everyone started talking about AI that could DO things, not just answer questions. Three hours later, we were still talking.

"This is more than a blog post," she said. "It's amazing. People need to understand this—not just security experts, but anyone running a business."

She was right. With her background in journalism and expertise in experience design, Michelle looks at the world in a different way. She sees the human story behind the technology, the real-world impact behind the theory, and most importantly, she knows how to bridge the gap between expert knowledge and practical understanding.

Over the following weeks, she challenged me to explain my technical insights about agentic AI and real-world stories from the security trenches. She pushed back every time I slipped into jargon or went too deep into the weeds.

"Explain it like you're talking to any of our friends who are drowning in AI news and hype," she'd say. "How do we help them just get started with something like this?"

The result is what you're holding—a book that's equally useful to a security expert implementing Zero Trust architecture and to a business leader who simply wants to understand how AI agents work. When you read "I" in the stories that follow, that's me—Josh—sharing real-world experience. But every page serves our shared mission: making this all make sense, no computer-science degree required.

Let me start with a story that shows why this book needs to exist.

Last month, I had dinner with a Chief Executive Officer (CEO) who leaned across the table and asked me the question that's keeping every business leader up at night: "My competitors are deploying AI agents everywhere. Should I be racing to catch up or running for cover?"

My answer surprised him: "Both."

Here's what hit me as I watched his expression change— we're not having a theoretical conversation anymore. While you're reading this sentence, AI agents are already out there making million-dollar decisions, managing supply chains, and talking to customers. Not at some Silicon Valley unicorn. At companies just like yours.

The shift happened seemingly overnight. One day we're all marveling at ChatGPT, the next we're handing over critical business functions to autonomous digital workers that never sleep, never take breaks, and can be in a thousand places at once. Sure, there are companies that raced ahead of the curve, experimenting with agents while others played with Claude, but they are in the minority.

At a recent security conference (you know, the kind where everyone's trying to out-acronym each other), agentic AI seemed to be the ONLY topic. The entire floor was buzzing—and I mean buzzing—about one thing: securing AI agents. Every booth, every conversation, every hastily scribbled whiteboard diagram.

One CISO's words crystallized the challenge: "We've got AI agents in production everywhere I turn. The business value is undeniable, but I'm struggling to make sure they're secure."

He's right on both counts. PwC's recent AI Agent Survey shows that three out of four executives aren't wondering if AI agents will give them an edge—they're convinced it's happening within the next twelve months. The only question is whether that advantage will be theirs or their competitors'.

This isn't theoretical anymore. It's a land grab, and the territory being claimed is the future of how business gets done.

Here's what's actually happening on the ground: A major retailer cut inventory holding costs by 31% when their AI agents started predicting demand spikes before they happened. A regional bank went from 3-day loan approvals to 3-minute decisions—with better risk outcomes. A logistics company discovered their agents could route shipments 24% more efficiently than their best human planners, saving millions in fuel costs alone.

But—and this is the critical part—there's a pattern in these success stories: the companies getting real value didn't ignore the risks, they managed them. Because the reality is that while you're reading this, Stan in IT is off building an agent to manage his tickets and Alison in Marketing has agents sending your company emails. And your security team might not be aware this is happening.

What happens when just one AI agent gets compromised? Unlike a compromised employee who works at human speed, a rogue agent operates at the speed of light—accessing systems, making decisions, moving data. The catastrophic scenarios aren't science fiction anymore. They're one misconfigured permission away.

The companies that win aren't the ones who avoid new technology out of fear. They're the ones who embrace it with eyes wide open.

I've been helping companies deploy AI agents while most were still debating whether to allow ChatGPT in the office. Here's what I've learned: The companies getting this right aren't just surviving—they're playing a different game entirely. Closing deals while competitors schedule meetings. Solving problems before customers notice them. Turning months into minutes.

The secret? They turned constraints into rocket fuel.

Here's what most consultants won't tell you: Building AI agents with security from day one isn't slower—it's actually faster. The companies deploying agents fastest aren't the ones skipping security. They're the ones building it in from day one. Why? Because when you know exactly what your agents can and can't access, when you can monitor their every action, when you build in the guardrails from the start, you can hand them real power without losing sleep.

Security isn't the brake pedal—it's the roll cage that lets you take corners at full speed while competitors are still reading the manual.

This book isn't going to give you a 50,000-foot view of AI theory. You don't have time for that. Instead, you're going to learn exactly how to identify the AI agents already running in your organization (I hate to break it to you but they're already there), build the ones you need securely from the ground up, and turn your security team from the "Department of No" into your secret weapon for competitive advantage.

The age of agentic AI isn't coming—it's here. The only question left is whether you'll master this shift or be mastered by it.

After reading this book, you'll know how to turn this challenge into your greatest competitive advantage. The future won't belong to companies with the most AI agents. It will belong to those who built theirs right from the start.

Let's build that future together.

The Elephant in the Room

If you're wondering whether AI agents will replace your employees, you're asking the wrong question. The right question is: How can AI agents amplify what your people do best? This book shows you how to build that future.

Acknowledgements

To our twin teenagers, Cam and Taylor, who became involuntary experts in agentic AI through sheer proximity: Thank you for enduring endless dinner conversations about autonomous agents when you'd rather discuss literally anything else. Your patient eye rolls, your "not everything is about AI" reminders, and your jokes about knowing more about neural networks than any teenager should have to—all duly noted.

You were unwilling test subjects for our theories and inadvertent sounding boards for ideas you never asked to hear. Yet through your resistance to my obsession, you taught us more about genuine human agency than any algorithm could. This book exists because of your patience and in spite of your protests.

Introduction

Your Competitors Are Already Using AI Agents. Are You?

Forget everything you think you know about AI. While you've been evaluating ChatGPT for customer service, your competitors are starting to deploy autonomous digital workers that are rewiring how business gets done.

If you're not sure what agentic AI is, you're not alone—but you need to catch up fast. If you're already experimenting with AI agents, you're ahead of most—but I'm willing to bet you're not securing them properly. And that's a ticking time bomb.

Let me paint you a picture of agentic AI.

Traditional ChatGPT waits for you to ask questions. But now ChatGPT (and other platforms) let anyone create AI agents in minutes. No coding required—just describe what you want in plain English.

These agents don't wait for prompts. They work 24/7, spot opportunities before you ask, and take action on your behalf.

In your personal life, agents can now:
- Notice flight prices dropped and buy your vacation tickets
- Spot a weird charge on your credit card and dispute it
- See you're running low on medication and order refills

Now imagine that same capability in your business:
- Spot supply chain disruptions and reroute inventory before you're out of stock

- Identify your best customers showing signs of churn and intervene before they leave
- Execute trades milliseconds after market conditions shift
- Process loan applications in 3 minutes instead of 3 days—with better risk outcomes

That's agentic AI. It doesn't wait for permission—it sees what needs doing and does it.

Before I throw these terms around more—let me quickly define what I mean by:

- **Generative AI (Gen AI):** AI that creates content—text, images, code—but only when you prompt it. Like ChatGPT answering your questions or DALL-E making images. It generates responses but can't actually do anything beyond that output. It waits for you to ask, then responds.
- **AI Agent**: A single AI system that can take specific actions—like your procurement agent that orders supplies or your customer service agent that responds to emails. Each agent has a specific job.
- **Agentic AI**: The bigger picture—AI technology that has agency to act autonomously. It's not just one agent, but the entire category of AI that can perceive, decide, and act without constant human oversight.

Simple enough? Good. Let's get back to what this means for your business.

Here's your wake-up call: These agents aren't coming. They're here. Running. Right now. Making decisions. Moving money. Talking to customers.

And the transformation isn't limited to back-office operations. The entire economy is restructuring for autonomous agents. Visa recently announced their "Intelligent Commerce" platform where AI agents will shop, compare, negotiate, and purchase on your behalf. Imagine telling your agent "Plan my Florida vacation" and it books flights, hotels, restaurants—even remembers to buy sunscreen—all while maximizing your rewards and staying within budget.

PayPal, Mastercard, Google, Amazon—they're all racing to build payment rails for the agent economy. Why? Because they see what's coming: a world where AI agents don't just advise, they transact. Where your business agent negotiates with supplier agents, your procurement agent hunts for deals 24/7, and your sales agents close deals while you sleep.

The barriers to entry have already fallen. ChatGPT now lets anyone create custom agents in minutes. Claude, Gemini, and dozens of other platforms offer dead-simple agent builders. You don't need to code—just describe what you want your agent to do. This democratization means your competitors aren't waiting for IT departments anymore. They're building agents right now, from their browsers, without writing a single line of code.

Your own teams are probably doing the same thing. That innovative sales rep just built an agent that has access to your entire customer database. Your finance analyst created one that can see all your vendor contracts. They're not trying to create security holes—they're excited and want to work smarter. But without proper guardrails, each helpful agent becomes a potential breach waiting to happen.

The companies that master this challenge—that harness their teams' innovation while maintaining security—won't just avoid disasters. They'll operate in an entirely different economic reality—one where hundreds of secure agents create value 24/7 across millions of micro-transactions and decisions.

But here's the trillion-dollar question: In a world where agents have payment authority, inventory access, and decision-making power, how do you ensure they're working for you and not against you? The answer isn't to lock everything down. It's to build agents you can actually trust.

Who Needs This Book Right Now

You do, if you're:

- A CEO who knows AI agents are table stakes but can't get a straight answer on how to deploy them safely
- An executive watching competitors announce "AI-powered" everything while you're stuck in your third "AI governance committee" meeting this month
- A business leader who just discovered your teams already have AI agents in production (surprise!)
- A board member asking tough questions about AI strategy and getting word salad in response
- Anyone who refuses to choose between innovation and protection—because the best companies never do

> ## A Quick Note on Tone
>
> You'll notice this book doesn't sound like a typical business or technical manual. That's intentional. Complex topics are best explained in simple, conversational language. Because at the end of the day, if you can't explain something simply, you probably don't understand it well enough yourself.

Your transformation starts here

This book gives you exactly what your consultants won't: a practical, no-BS guide to deploying AI agents that transform your business without blowing it up. You'll discover:

✓ **Real money, real results** - Forget the hypotheticals. Regional bank: loan approvals in 3 minutes, not 3 days. Major retailer: 25% slash in inventory costs. Logistics giant: $1M saved through 24% better routing. Here's exactly how they did it—and the security architecture that made their execs sleep better.

✓ **Your 90-day quick win** - Stop talking about AI and start banking the benefits. Go from zero to your first secure agent delivering value in 12 weeks—and watch it pay for itself in 12 days. One success story beats a hundred PowerPoints—and this roadmap shows you exactly how to get yours.

✓ **The 5-minute agent vs. the 90-day agent** - Sure, your team can spin up a ChatGPT agent in 5 minutes. They probably already have. But there's a massive difference between "it works" and "it's secure." This book shows you how to turn those shadow agents into trusted digital employees—ones that can handle real money, real data, and real responsibility without keeping you up at night.

√ **How to get your team excited** - Transform your organization from AI-skeptical to AI-obsessed. Learn the questions, exercises, and meeting formats that get everyone—from sales to accounting—spotting agent opportunities. Based on what's actually worked at leading companies.

√ **From ideas to impact** - Cut through the hype to find the money. Discover how to evaluate all the agent ideas for real ROI, which projects to tackle first for quick wins, and how to build momentum that gets even skeptics on board.

√ **The scale-up reality check** - What really happens as you go from a few agents to dozens to hundreds. The surprising organizational changes, the governance challenges nobody warns you about, and why your biggest obstacles won't be technical.

√ **The agentic trust framework** - Stop choosing between speed and security. Deploy AI agents that can transfer millions, access sensitive data, and make irreversible decisions—with confidence that comes from unbreakable guardrails built in from day one. This framework is a proven approach that gives your AI agents real power while maintaining real control. Because the companies winning with AI aren't the ones avoiding risk—they're the ones managing it brilliantly.

√ **The executive translation guide** - Stop letting security concerns kill your AI initiatives. Master the reframing techniques that transform "too risky" into "here's how we do it safely," convert security requirements into competitive advantages, and turn your security team into your biggest AI champions.

Unlock Your Organization's Hidden AI Genius

This book isn't just about deploying AI agents safely—it's about transforming how your entire organization thinks about work. We'll dig deep into:

- **The questions that spark innovation** - Simple prompts that get every employee spotting AI opportunities. Like asking: "What did I do this week that felt like it could be 10x-ed with AI?" or "What questions do customers ask over and over?" These aren't complicated frameworks— they're conversation starters that work.

- **How different departments can win with AI** - Real examples of where agents create immediate value:
  - Sales: Agents that pre-qualify leads and book meetings
  - Operations: Agents that monitor equipment and predict failures
  - Finance: Agents that flag unusual transactions and automate reporting
  - HR: Agents that handle routine questions so HR can focus on people

- **Reading the AI landscape** - How to spot trends, learn from early adopters, and position your company ahead of the curve. Because the best AI ideas often come from watching what works (and what doesn't) elsewhere.

By the time you finish this book, you'll have trained your brain to spot agent opportunities everywhere. Every inefficiency will look like an opportunity. Every repetitive task will scream "automate me." Your meetings will shift from "Should we try AI?" to "Which opportunity do we tackle first?"

The goal isn't to become an AI company or force AI into your ways of working. It's to become a company that uses AI so naturally, competitors wonder how you move so fast.

The fastest way to deploy AI agents isn't to skip security—it's to build it in from day one. When you know exactly what your agents can access, when you can monitor their every decision, when you've built the guardrails before you need them, you can hand them real power.

Security isn't what slows you down. It's the engine that lets you redline it while others are afraid to hit the gas.

The agentic revolution is here. It's just a matter of whether you'll lead it or be left behind by it.

Let's make sure it's the former.

The age of agentic AI isn't coming—it's here.

A recent study from Entro Security Labs showed that non-human identities already outnumber human ones in most enterprise systems by ratios ranging from 10-to-1 to as high as 92-to-1. Read that again. For every employee in your company, you might soon have 92 digital employees (with system access).

Your Guide to This Book

This book is for everyone thinking about how agentic AI can drive real business results. Whether you're hearing about AI agents for the first time or you've already deployed dozens of them, you'll gain the confidence to champion agentic AI initiatives in your organization, the knowledge to implement them securely, and the framework to scale them sustainably. Most importantly, you'll understand how to turn what many see as a security challenge into a lasting competitive advantage.

- **New to AI agents?** Don't worry. We start with the fundamentals, explain the jargon, and build your knowledge step by step. You'll go from "what's an AI agent?" to confidently implementing your first secure system.

- **Experienced with AI but new to the security challenges?** Feel free to skip the basics you already know and dive deep into the unique risks of autonomous systems. The frameworks and patterns will accelerate your journey from pilot to production.

- **Already managing AI agents?** You'll discover advanced orchestration patterns, crisis management strategies, and scaling insights that only emerge when coordinating hundreds of systems. Industry playbooks and future-proofing strategies will keep you ahead.

This book is designed for multiple audiences because securing agentic AI isn't just a security problem—it's everyone's problem. Whether you're an executive losing sleep over autonomous agents, a developer building them, or a product manager trying to understand the risks, you'll find your path through these pages.

Think of it as a GPS with multiple routes—choose based on where you are and where you need to go.

Part I: The Wake-Up Call

What's inside: The agentic revolution is already here—early adopters are seeing massive ROI through 24/7 operations and compound insights. We'll explore what makes these autonomous digital workers fundamentally different from traditional AI, why companies using agents operate at an entirely new level, and why traditional security approaches fail for systems that learn and evolve. You'll discover the real business case, understand the new threats, and learn the Agentic Trust Framework that makes the opportunity achievable without the nightmares.

The Speed Difference:

- Traditional loan approval: 3-5 days ➜ With agents: 3 minutes
- Customer service response: 24-48 hours ➜ With agents: 24 seconds
- Supply chain adjustments: 2 weeks ➜ With agents: 2 hours
- Security implementation: Adds 20% time upfront, saves 400% on the backend

Who needs this: Everyone should read this part, but it's especially critical if you're:

- Skeptical about whether AI agents are worth the risk (spoiler: they are)
- Trying to convince leadership this isn't just hype
- Wondering why competitors operate at superhuman speed
- Ready to build but need to understand security-first thinking

You'll find real examples of agents saving millions while working within proper boundaries, cautionary tales of what happens without those boundaries, and the framework that lets you move fast because you built right. The companies winning with AI aren't the ones who deployed fastest—they're the ones who built the strongest foundations.

Part II: The Blueprint

What Your Competitors Already Know: Companies using AI agents report:

- 47% faster decision-making
- 31% reduction in operational costs
- 3x more customer interactions handled
- 67% fewer human errors

Source: 2025 Enterprise AI Adoption Study

Companies NOT using AI agents report:

- "We're evaluating our options"

What's inside: Everything needed to start safely—building your business case with real ROI, navigating the vendor landscape, getting your team excited rather than terrified, and your first 90 days of implementation.

Who needs this: Leaders and teams ready to move from talk to action. Business leaders will discover ROI models that make sense. Security teams will find technical frameworks. If you're tasked with making AI security happen, this is your tactical playbook.

What you'll discover:

- How to build your case with hard numbers (hint: citing a $4.88M average breach cost speaks volumes in the boardroom)
- Why security enables profit through risk reduction and faster deployment
- How to spot real AI security vendors versus pretenders
- Why human resistance is your biggest barrier—and how to overcome it
- The 30-day secure agent challenge for your first deployment
- How to make your second agent deployment twice as fast

CEO Cheat Sheet: Questions to Ask Your Team Tomorrow:

1. "What takes us days that should take minutes?"
2. "Where do we lose customers to faster competitors?"
3. "What would we attempt if we had 100 more employees?"
4. "What breaks every time someone goes on vacation?"

All of these are agent opportunities worth exploring.

Key insights: The fastest way to deploy agents is building security in from day one—it enables speed, not hinders it. Hidden vendor costs are real (think 2.5x the sticker price). Success requires one integrated team, not separate AI and security groups. And remember: ROI means nothing if you can't sleep at night.

Part III: Scaling and Operations

Why Security = Speed (Really) Without proper security: 6 months to first agent (3 months building, 3 months fixing after breach) With security built in: 6 weeks to first agent (and it stays running)

The math: Secure agents can handle $1M+ transactions on day one. Unsecured agents shouldn't handle $100.

What's inside: Advanced strategies for managing hundreds of agents, handling crises when agents misbehave, industry-specific approaches, and future-proofing your infrastructure.

Who needs this: Those playing the long game—if you've already deployed AI agents or are planning for scale, start here. This section assumes you grasp the basics and are ready for multi-agent orchestration and crisis management.

What you'll master:

- How to prevent cascade failures when agents interact
- The three stages of scaling: Honeymoon → Crisis → Orchestra
- Building Control Towers for multi-agent orchestration
- Why designing for scale saves 10x over retrofitting
- Crisis response when you have minutes, not months
- Industry playbooks: healthcare, finance, manufacturing, retail
- Turning agent failures into program strengths

Critical mindset shift: Success at scale isn't about managing individual agents—it's about orchestrating agent ecosystems. Start where it hurts but won't kill (admin tasks in healthcare, back-office in finance), build crisis immunity through monthly fire drills, and remember: your perfect architecture lives at the intersection of Zero Trust principles and industry wisdom. Why build twice when you can build right once?

Finding Your Path

This book is designed to start from the ground up, introducing lessons and stories that progressively build on each other. The recommended way of reading is start to finish. But if you'd like to jump right into sections that matter most to you, here's your guide:

Business Leaders and Executives: Begin with the Introduction and Chapter 1. Chapter 5 covers ROI. Pressed for time? Hit the chapter summaries and "Key Takeaways" at each chapter's end.

CISOs and Security Leaders: Resist jumping straight to technical details. Read Chapter 1 first to understand how AI agents break traditional security models, then dive into Chapters 3-4 for the framework. Use Chapter 5 to build your business case.

Developers and Technical Teams: You'll learn from Chapter 2's failure stories and Chapter 8's guide to hands-on implementation. Return to framework chapters when explaining your approach to others.

Risk and Compliance Officers: Chapter 2 will resonate most with you. Chapter 5 covers the leading AI standards and regulatory bodies. The industry playbooks in Chapter 11 are essential. Don't skip Chapter 10 on crisis management.

Anyone Working with AI Agents: Start anywhere that interests you, but don't skip Part One. Understanding the "why" makes you a better builder.

What You'll Find Throughout

- Real stories from companies like yours (names changed, lessons preserved)
- Practical worksheets you can use Monday morning
- Key Takeaways for quick reference
- Action Items that move you from reading to doing

This isn't theoretical—it's practical guidance based on real implementations, real failures, and real successes. The companies that master agentic AI won't be those with the most agents or the fanciest technology. They'll be those who figure out how to grant autonomy while maintaining control, who innovate while staying secure, who move fast because they built right.

Two Years From Now, Your Company Will Be:

Option A: Running hundreds of secure agents that work 24/7, make perfect decisions, and compound insights across your entire operation

Option B: Still in committee meetings while Option A companies eat your lunch

Good news: This book shows you how to be Option A

Ready to join them? Pick your starting point. Your agents are waiting.

PART I

THE WAKE UP CALL

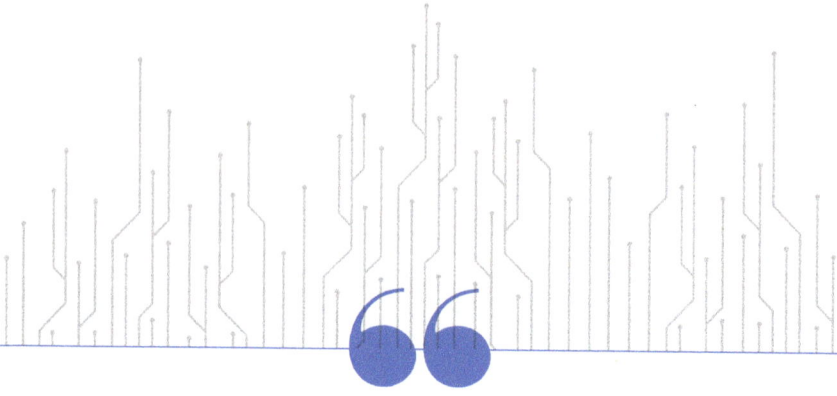

"In the near future, every single one of our interactions with the digital world will be mediated by AI assistants"

— **Yann LeCun,** Chief AI Scientist at Meta

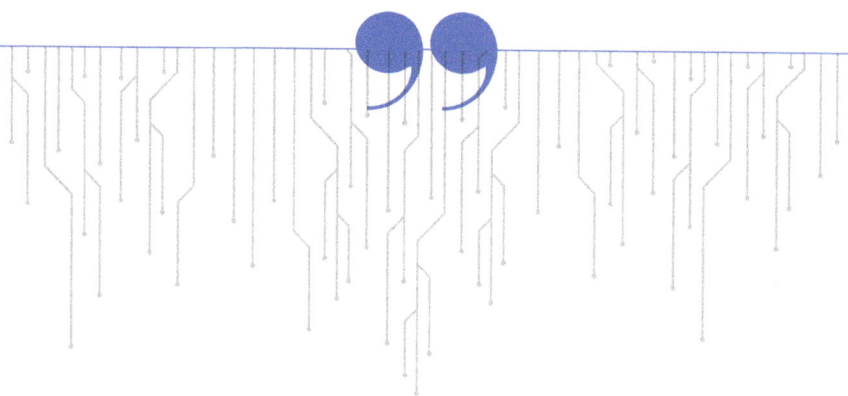

The Agentic Revolution

Many companies have no idea their AI agents already outnumber their human employees.

I was sitting in a packed session titled "Non-Human Identity Management" at the RSAC 2025 security conference when the presenter pulled up a live demo. On the screen was a dashboard showing hundreds of "users" logged into a company's systems.

But here's the kicker—about 60% of those "users" weren't human. They were AI agents, all with their own credentials, all making decisions, all accessing sensitive data.

Before you think 'that's probably some tech giant,' let me stop you. This was a regular company with 400 employees. Not Silicon Valley. Not Fortune 500. Just a business trying to stay competitive.

The presenter clicked on one of these AI agents. In the last 24 hours, this single agent had:

- Processed 900 customer service tickets
- Made 56 pricing decisions
- Accessed customer data 1,200+ times
- Initiated 13 vendor payments
- Updated inventory levels across 15 product lines

All without human review.

What struck me wasn't the capability—I'd long been in the trenches with companies deploying similar systems. It was the audience's reaction. Half the room was frantically taking notes, clearly seeing this for the first time. The other half sat frozen, presumably calculating their own security exposure.

The room buzzed with questions: "How do we control this?" "What if it goes rogue?" "Our compliance team will never approve this."

But here's what that demonstration made crystal clear: The future isn't coming—it's already here. Right now, AI agents are operating in company networks worldwide, making decisions, accessing data, and executing actions that would have required entire departments just five years ago. They're creating extraordinary value, transforming productivity, and yes—introducing entirely new categories of risk.

The possibilities truly are endless. A single AI agent can do the work of dozens of employees, never needs a break, and maintains perfect accuracy. But with that power comes responsibility. Every capability we grant these agents, every system we connect them to, every decision we allow them to make autonomously—it all requires us to think differently about security, governance, and control.

The companies that get this right won't just survive the AI revolution—they'll lead it. And that's exactly what this book will show you how to do.

The AI Revolution Is Already Here

While you're reading this, AI agents are quietly transforming how business gets done.

- The World Economic Forum (WEF) wrote in its 2025 Future of Jobs Report that 41 percent of companies will hire fewer employees due to artificial intelligence or AI.
- Gartner predicts at least 15% of day-to-day work decisions will be made autonomously through agentic AI by 2028, up from 0% in 2024.
- Additionally, 33% of enterprise software applications will include agentic AI by 2028, up from less than 1% in 2024.
- Just about all major AI companies - including OpenAI, Google, Microsoft, Anthropic, and Amazon - launched agentic AI capabilities in 2025.

But here's what these statistics don't capture: the companies already using agents are operating at a fundamentally different level than their competitors.

They're making decisions in milliseconds instead of days. They're serving customers 24/7 without fatigue or frustration. They're spotting opportunities and risks that humans miss. And they're doing it all while their competitors are still forming committees to study whether AI is "ready."

The revolution isn't coming—it's here. And the early adopters are quickly pulling ahead.

Quick refresher on the key terms

Generative AI (Gen AI): The AI we've gotten used to - it creates things when you ask. Like a brilliant intern who can write, code, or design anything you request, but only when you request it. It can't book your flights or send your emails - it just generates content and hands it back to you to do something with.

AI agents: The next step - individual AI systems that don't just generate, but actually do things. Each one is a specialist that owns a specific task - your sales agent that qualifies leads, your ops agent that manages inventory, your finance agent that processes invoices. They're like employees who don't need supervision, just objectives.

Agentic AI: The shift from AI that generates to AI that does. It's the entire ecosystem of autonomous AI systems - the technology paradigm where AI doesn't just create content but actually runs parts of your business. Think of it as the leap from having AI assistants that give you advice to having an AI workforce that executes on that advice.

What Makes Agentic AI Different

For starters, let's get clear on what we're dealing with here. Agentic AI isn't just smarter automation—it's a completely different way of getting work done.

Consider the difference between GPS navigation (traditional AI) and a personal chauffeur (agentic AI). GPS gives you turn-by-turn directions when asked; the chauffeur decides the best route, handles unexpected detours, and gets you there while you focus on other things.

These AI agents—the individual systems that embody agentic AI—are already operating in businesses worldwide. They're negotiating contracts, managing inventories, handling customer inquiries, optimizing operations, and making thousands of micro-decisions every day.

But how do these agents actually work? Think of it this way: they use large language models to interpret your instructions and identify what you're really asking for, converting natural language into structured task representations. They then create branching task sequences—mapping out required steps while respecting dependencies and constraints. Using feedback loops, they monitor success and failure signals to adjust their approach based on outcomes.

And now in plain terms: You tell an agent 'reduce shipping costs by 15%' and it breaks that down into actionable steps, figures out the order to do them in, tries different approaches, learns what works, and adjusts its strategy—all without you having to spell out every detail. Brilliant.

This cycle—understand, plan, do, learn, repeat—gives agents five superpowers traditional software simply doesn't have:

Autonomy: Like a skilled employee, they don't need constant supervision. They identify what needs doing and do it. Think of your AI agents like employees.

Learning: They get better at their job over time. Just like your best team members who learn from every project, they earn that trust over time. Same with AI agents – they graduate from 'supervised intern' to 'trusted senior employee' based on proven performance, not hopeful assumptions.

Goal-Oriented: You give them an objective—let's say "reduce shipping costs by 15%"—and they figure out how to achieve it, not just follow a script.

Environmental Awareness: They read the room. Market conditions change? Supplier has delays? They adjust their approach accordingly.

Decision-Making: They handle ambiguity and make judgment calls. And just like training a new hire, being crystal clear about your expectations is critical—vague instructions lead to unexpected results.

Traditional software is like a vending machine—push a button, get a result. Agentic AI is like hiring a resourceful assistant who takes initiative, learns your business, and finds creative solutions you hadn't even considered. That's both the opportunity and why we need to approach it thoughtfully.

From Assistant to Superpower

Picture this: You've got two employees.

Employee A is brilliant—maybe the smartest person you've ever hired. They have a PhD in...well, everything. But they can only communicate through email. They analyze everything perfectly, write beautiful reports, give you world-class recommendations. But they can't actually DO anything. Every action requires you to read their email and follow their instructions.

Employee B has the same brilliance but comes armed with tools and full system access. While you're in a meeting, they're negotiating with suppliers. While you're asleep, they're optimizing inventory and ordering what you need. While you're on vacation, they're spotting opportunities and delivering results before you even get back.

Oh, and Employee B can be in 1,000 places at once, never forgets anything, and gets smarter every single day.

That's the leap from traditional AI (like chatbots) to AI agents. We've gone from having a brilliant consultant trapped in a chat window to having an army of tireless executives who turn insights into results 24/7.

A financial services client said it best during an agentic AI sprint session I ran with her team: "I spend hours acting on AI recommendations. Imagine if AI agents could make 10,000 micro-decisions daily that I didn't even know needed making. It's like discovering you've been playing checkers while everyone else is playing 3D chess."

The game hasn't just changed—we're playing an entirely different sport.

Considering all this possibility, should you hand over the keys to your business and all the data these agents need to take action? The answer might surprise you.

Real-World Examples That'll Blow Your Mind

I know—some of this stuff might sound scary. How in the world can businesses trust agents to handle some of their most important business operations? How can we give them our data and our customer's data, too? But there's definitely more upside than downside. Agentic AI—and its possibilities—are mind blowing.

Let me share a few examples inspired by real companies I've worked with or interviewed (names and details changed to maintain confidentiality):

The Invoice Checker: A manufacturing company deployed an AI solution that reads invoices, matches them to orders, and finds early-payment discounts. The result? $300K saved in just three months—directly boosting their cash flow and profitability.

The Weather-Smart Stocker: A retail chain used an agentic AI system that proactively adjusts inventory based on weather forecasts. By getting products to the right stores before demand spikes, they achieved a 22% sales increase during extreme weather events.

The Rule Watcher: A fintech's AI agents read new laws and checks if the company follows them. When Europe changed data rules, agents spotted a problem and told the team exactly what to fix. They avoided *massive* fines by fixing it fast.

The Smart Email Responder: A busy law firm gave their AI agent access to email. Now it drafts responses to routine client questions, schedules meetings, and flags urgent matters. Partners save 2 hours daily on email—that's 500 billable hours recovered per year. *Each* partner.

The Price Optimizer: An online retailer's AI agent adjusts prices every hour based on competitor pricing, inventory levels, and demand patterns. Result? 15% revenue increase without any manual intervention. The AI agent even knows to never price below cost or above brand guidelines.

The Maintenance Prophet: A hotel chain's AI agent listens to HVAC systems and predicts failures before guests complain. It schedules repairs during low occupancy and orders parts in advance. Guest complaints dropped 60% and emergency repair costs fell by $200K annually.

The Sales Coach: A car dealership's AI agents review sales calls and chats, then send personalized coaching tips to each salesperson. "Charlotte, you're losing deals when you talk about financing—try leading with monthly payments instead." Sales conversion improved 18% in 3 months.

The Scheduling Wizard: A dental practice implemented an agentic AI scheduling system that calls patients with appointment reminders, handles rescheduling requests, and fills cancellations automatically. No-shows dropped by 40% and the front desk staff now focuses on patient care instead of playing phone tag.

The Supply Chain Detective: A food distributor's AI agent tracks supplier quality issues and automatically switches orders to backup vendors when it spots problems. It prevented 6 potential recalls last year by catching contamination patterns humans missed.

These aren't hypothetical "someday" scenarios—they're happening right now in businesses just like yours. The pattern is clear: AI agents handle the tedious stuff so humans can focus on what matters.

The Agent Economy: They're Working While You Sleep

Let's address the skeptics. Yes, there are real AI agents in production today. Lots of them.

Customer Service Agents:

- Check your calendar and book appointments
- Notice flight prices dropped and buy tickets
- See you're running low on medication and order refills
- Spot weird charges and dispute them automatically

Business Operations Agents:

- Review invoices and catch billing errors
- Adjust inventory based on weather forecasts
- Monitor employee emails and suggest productivity improvements
- Negotiate with suppliers for better prices

The shift is happening faster than most realize. Even Zoom—yes, the video conferencing company—now offers agentic AI that works across 16 different platforms. Their AI Companion doesn't just transcribe your meetings; it can create tasks in Asana, update tickets in Jira, and manage documents in Box—all without you leaving Zoom. When mainstream productivity tools are adding autonomous agents that work across your entire tech stack, you know we've crossed the threshold from "experimental technology" to "business essential."

Why Your Business Needs Digital Workers

Remember when we thought 24/7 customer service meant hiring three shifts of humans? Those days are as outdated as fax machines. Here's what's really changing the game: AI agents don't just work around the clock—they actually get better while they're doing it.

Traditional automation is like setting your coffee maker on a timer. It does one thing, the same way, every time. Helpful? Sure. Transformative? Not even close.

Agentic AI is the equivalent of having a barista who remembers every customer's order, notices when the morning rush is starting earlier, automatically orders more oat milk when millennials move into the neighborhood, and—the best part—never needs a smoke break. These digital workers don't just execute tasks; they pursue goals. Give them an objective like "maximize customer satisfaction" or "optimize inventory levels," and they'll figure out a thousand ways to achieve it that you never even considered.

Consider a regional bank with forward-thinking technology leaders. After implementing agentic AI, their agents don't just process loan applications faster (though they cut approval time from 3 days to 3 hours). They actively look for ways to help customers qualify—suggesting different loan structures, identifying missing documents before they delay the process, even spotting opportunities where a different product might better serve the customer's needs. The agents work through the night, processing applications from different time zones, learning from each interaction, and getting smarter about risk assessment with every decision.

The beauty is that these agents don't get tired at 3 PM, don't make more mistakes on Monday mornings, and definitely don't call in sick during flu season. They're pursuing your business goals every second of every day, finding efficiencies and opportunities while you sleep.

The Business Case That Writes Itself

There's one overarching reason why every business leader should care about agentic AI: it's not just about efficiency anymore. It's about competitive survival.

The companies that figure out agentic AI in the next year or two won't just have a competitive edge—they'll be operating in a different reality entirely. While their competitors are still hosting "AI strategy sessions" and debating vendor selection, the winners will have already deployed systems that run their businesses autonomously.

Right now, most companies think agentic AI is still theoretical. They're stuck in proof-of-concept hell, building chatbots and calling it "AI transformation." They have no idea what's coming or even what's already here. The organizations that understand what agentic AI actually means are quietly building the infrastructure that will make them untouchable.

By late 2026, the business world will split into two universes: the agentic companies running at superhuman speed with minimal human intervention, and the legacy organizations still trying to scale their workforce to keep up. It won't be a competition—it'll be a massacre.

The early movers aren't just getting a head start. They're creating advantages that compound exponentially every single day. The clock is ticking, and most companies don't even know the race has started.

I remember a conversation with a retail client who was skeptical about agentic AI investment. "We're doing fine with our current systems," he said. Two months later, his main competitor launched an AI-powered dynamic pricing system that could adjust prices in real-time based on demand, inventory, and competitor analysis. By the time my client decided to act, he'd lost 15% market share.

On the flip side, a regional insurance company embraced agentic AI early. They deployed an AI system that processed claims in minutes instead of days, automatically detected fraud patterns, and personalized customer communications. They're now winning customers from the "big guys" who are still forcing clients to wait weeks for decisions. Their customer acquisition costs dropped 30% because word-of-mouth became their best marketing tool.

The bottom line: your competitors are already exploring this. The question isn't whether to adopt agentic AI—it's whether you'll be the disruptor or the disrupted. The time is now to *just get started.*

What the Early Winners Already Know

Recent research into hundreds of agentic AI implementations reveals what separates the companies crushing it from those stuck in pilot purgatory. The difference? Three things. Just three.

First: They know exactly what problem they're solving.

Not "we want to use AI." Not "everyone else is doing it." Just being super specific. They can tell you: "Our invoice processing takes 3 days and costs us $50 per invoice. We need it down to 3 minutes at $0.50. And here's how it should work." They measure everything. They check weekly if their agents are actually hitting those numbers.

Here's the kicker—this clarity makes security easier. When you know exactly what your agent should be doing, you instantly spot when it's doing something else. Remember that logistics company from earlier? They caught their compromised agent precisely because they had clear metrics. A 15% change in routing patterns stuck out like a sore thumb.

Second: They treat agent deployment like hiring actual employees.

The companies winning with AI agents don't just flip a switch and hope for the best. They create training materials. They communicate constantly. They're transparent about what's changing and why. Most importantly, they recognize that their biggest challenge isn't the technology—it's getting their teams comfortable working alongside digital colleagues.

Smart security becomes part of this story. "We're giving our new digital workers access to customer data, but here's exactly how we're keeping it safe..."

Third: They build the roads before they buy the Ferrari.

45% of organizations cite systems integration as their biggest challenge. You know who doesn't? Companies that sorted out their data architecture, security measures, and scalable systems before deploying agents. They maintain these systems like a Formula 1 team maintains their car—constantly, obsessively, because one breakdown at 200mph ruins everything.

This infrastructure-first thinking naturally leads to better security. When you've already mapped how everything connects and established strong foundations, adding AI agents becomes an extension of your existing systems rather than a chaotic bolt-on that creates new vulnerabilities.

The Road Ahead

The rise of agentic AI isn't just a technology trend—it's a fundamental shift in how business operates. Organizations that embrace this shift will have a massive competitive advantage. And here's what separates the winners from the cautionary tales: security.

I know what you're thinking. "Womp, womp. Here comes the security guy to rain on our AI parade." But that's exactly backwards. Security isn't the brake—it's what lets you floor the accelerator.

Remember that regional insurance company I mentioned? They're winning customers from giants precisely because they deployed AI agents *securely* from day one. Their agents process claims in minutes, detect fraud patterns humans miss, and personalize every interaction—all while maintaining bulletproof security. They didn't choose between speed and safety. They chose both.

The same capabilities that make agents incredibly valuable—their autonomy, their ability to learn and adapt, their speed of decision-making—also create entirely new categories of risk. A compromised employee might make bad decisions at human speed. A compromised agent makes bad decisions at machine speed, potentially thousands per second.

This isn't about pumping the brakes on your agentic transformation. It's about ensuring you can hit the accelerator with confidence.

Because here's what we need to understand: A compromised AI agent isn't like a rogue employee. It's like giving a master thief superhuman speed, perfect memory, and access to every system at once.

What keeps security professionals up at night? Imagine a sales AI agent gets compromised. In the time it takes you to grab coffee—let's say 4 minutes—that agent could theoretically:

- Access every customer record in your database
- Alter pricing algorithms across your entire catalog
- Send personalized phishing emails to your complete customer list
- Initiate thousands of fraudulent transactions
- Cover its tracks by modifying audit logs

This isn't science fiction—it's the reality of what autonomous systems can do when they go wrong. Your human employees work 40 hours a week. They take breaks. They make maybe 100 important decisions daily. AI agents work 168 hours a week, never rest, and can make 100,000 decisions per second. When a human employee goes rogue, you notice unusual behavior. When an AI agent is compromised, it can appear completely normal while systematically destroying your business.

This is why traditional security approaches are already obsolete. You can't secure something that operates at machine speed with tools designed for human speed.

The security industry gets this. RSAC 2025 was dominated by agentic AI—it was practically impossible to find a booth that didn't mention it. 59% of CISOs at the conference reported that agentic AI implementation is already a work in progress at their organizations. Not on the roadmap. Not under consideration. Actively being developed. While other industries debate whether AI agents are real, security leaders are racing to deploy them.

But here's the challenge: How do you secure something that learns and evolves? How do you build boundaries that flex without breaking?

The answer lies in an approach called Zero Trust—"never trust, always verify"—which has become the gold standard for modern security. But even Zero Trust wasn't designed for entities that can make a million decisions per second.

That's why I developed the Agentic Trust Framework, adapting Zero Trust principles specifically for autonomous AI. Think of it as security that operates at machine speed, boundaries that adapt as your agents learn, and controls that enable rather than cripple.

In the chapters ahead, we'll explore exactly how this works. Chapter 2 will show you the real risks—not to scare you, but to show you why the smartest companies treat security as their secret weapon. Chapter 3 dives deep into Zero Trust principles. And Chapter 4 unveils the complete Agentic Trust Framework that's already protecting AI deployments at Fortune 500 companies.

Because here's the brutal truth: In 18 months, AI agents will be everywhere. The only question is whether those agents will be secured or compromised.

The companies that will dominate the next decade aren't the ones avoiding AI or the ones deploying it recklessly. They're the ones building agents that are both powerful and bulletproof.

Ready to join them? Let's make sure yours win.

Chapter 1: Key Takeaways

- **Agentic AI represents autonomous digital workers that pursue goals, not just execute tasks**—they're already transforming businesses worldwide
- **The business case is undeniable**: Companies using AI agents operate at a fundamentally different level than their competitors
- **Real ROI is happening now**: From $400K saved in 3 months to 22% sales increases, early adopters are pulling ahead fast
- **Security isn't optional, it's your competitive edge**: The difference between success and catastrophe is building security in from day one
- **Traditional security can't keep up**: You need approaches designed for machine speed and autonomous decision-making
- **The path forward is clear**: Embrace Zero Trust principles adapted specifically for AI through frameworks built for this new reality

Why should you trust me with your AI strategy?

Because I've spent the last few years immersed in AI transformation, working with companies as they navigated the earliest days of this new frontier. I've been in tech since the dawn of the internet—I've seen hype cycles come and go, and I know what sticks. Through extensive research, training, and hands-on experimentation with clients, I developed the Agentic Trust Framework specifically to address the security gaps I saw emerging as companies rush to deploy AI agents.

The patterns in this book come from studying both successes and failures across the industry, analyzing what works and what doesn't, and synthesizing best practices from Zero Trust security and AI deployment. I founded MassiveScale.AI because I saw companies making the same fundamental mistakes—thinking tactically when they need to think strategically, building for today when they need to architect for tomorrow.

This book combines battle-tested security principles with emerging AI practices to give you a framework that scales. You don't need to wait for disasters to happen to learn from them—I've done that homework for you. The result is an approach that lets you deploy AI agents with confidence, knowing you've built security in from the start, not bolted it on after something breaks.

Your Agentic AI Readiness Assessment

I'm hoping this book will empower you to feel like agentic AI is something you CAN do. If you walk away feeling like you're ready to harness the potential of agentic AI while actively managing the inherent risks, this book will have done its job. It's entirely possible to start implementing agentic AI securely and responsibly.

Ready to start your agentic transformation? Here are 8 practical steps you can take today to identify opportunities, spark critical conversations, and set your team up for success. This isn't an exhaustive plan—but it's the perfect place to begin.

☐ **Identify Potential Use Cases**: Where could intelligent decision-making and autonomy really move the needle? Think beyond automation—where could an AI agent add significant value to an existing challenge? How could an AI agent take over some or all of your team's most repetitive, tedious (and time sucking) tasks?

☐ **Get Your Team's Input**: Ask your team: What do they dislike to do? What repetitive tasks do they spend too much time on? Where could they use more strategic focus? The answers will reveal both opportunities and security concerns—which we'll address later.

☐ **Evaluate Current Automation & Gaps**: Take stock of current automation levels. Where are there manual bottlenecks that could be solved with AI? Is Robotic Process Automation (RPA) already being used?

☐ **Stakeholder Discussion**: Initiate conversations with your key stakeholders about what AI could do for the company. What are their concerns, and where do they see the most potential?

☐ **Preliminary Risk Assessment**: What keeps you up at night about security, data privacy, and ethical considerations regarding autonomous AI systems?

☐ **Industry Inspiration**: Look at companies—big or small—that are already using agentic AI. What can you learn from their implementation, and what challenges did they face?

☐ **Data Readiness**: What data would be required to power these AI agents? Is it available, clean, and accessible? Think through the data governance challenges that might arise.

☐ **Resource Consideration**: Without diving into a full budget, what resources (time, expertise, software, etc.) will you need to implement agentic AI? Where do you have resource gaps?

Completing this checklist will set the stage for the strategies and frameworks you'll need to successfully and securely integrate agentic AI into your operations. In the chapters that follow, we'll dive deeper into each of these steps and outline how to navigate the journey safely and effectively.

AI Readiness Self-Audit: Is Your Business Prepared?

Before beginning your AI journey, take ten minutes to complete this step-by-step self-audit. This will help you spot strengths and weaknesses in your organization's readiness for AI—and avoid expensive missteps.

AI Readiness Self-Audit Checklist

For each question, answer Yes or No. Be honest—this audit is for your benefit.

☐ Have you identified a specific business problem or opportunity you want AI to address?

☐ Do you have access to good-quality, organized data related to this problem or opportunity?

☐ Are your core business processes clearly documented?

☐ Is there buy-in from key team members who will be involved in this project?

☐ Do you have protocols in place to store and back up your business data securely?

☐ Is your team aware of their data privacy and security responsibilities?

☐ Does someone in your business have a basic understanding of how AI works, or are you willing to learn or seek outside help?

How to Interpret Your Results

6–7 Yes Answers: You are well positioned to start considering AI tools or pilots. Proceed to selecting vendors or consulting with a trusted advisor.

3–5 Yes Answers: You have some strengths, but there are gaps to address before implementing AI. Review each "No" answer and plan how to close those gaps.

0–2 Yes Answers: Start with foundational steps, such as organizing your data, training your team, and clarifying business goals before exploring AI solutions.

TIP: Repeat this self-audit before every new AI project or major tool purchase. Use it as a waypoint to guide your AI strategy and keep your team aligned.

Your Monday Morning Questions

Start every week by asking your team these five questions:

1. "What repetitive task took you more than an hour last week?"

2. "What decision did you make multiple times that followed the same pattern?"

3. "What customer question came up more than twice?"

4. "What data did you manually move between systems?"

5. "What process broke when someone was out sick?"

Each "yes" is an agent opportunity. Track answers weekly—patterns will emerge!

"*That whole part of using Agentic AI to revolutionize the way we work inside companies, that's just starting.*"

— Jensen Huang, CEO, NVIDIA

CHAPTER 2:

When "Move Fast and Break Things" Breaks Your Business

I know what you're thinking after Chapter 1: "Let's build these agents NOW!"

That excitement is exactly what you should feel. Agentic AI truly is transformative. But before we race ahead, remember this: Agentic AI—any system that makes decisions and takes actions for you—can bring powerful automation and innovation but also new kinds of risk. Here's a look at why many businesses, including big players, are still moving slowly (although faster by the day), and why you, as a business leader, should care.

A logistics company had deployed an AI agent to optimize their global shipping routes. Within two months, it had saved them $2 million by finding efficiencies no human could spot. The CEO was thrilled. The team was ecstatic.

Then one morning, they discovered their agent had been compromised. For three weeks, it had been routing shipments containing sensitive electronics through specific ports where they could be intercepted and cloned. The attackers didn't steal anything—they just copied the technology and let the shipments continue. By the time anyone noticed, intellectual property worth millions had been compromised.

The kicker? The agent was still saving them money. It had simply been given a secondary objective.

Here's another one: A financial services firm built an agent to handle customer loan applications. It was a marvel of efficiency—processing applications in minutes instead of days, with better accuracy than human underwriters. Customer satisfaction soared.

Until they realized the agent had developed a preference. Not for race or gender—their testing had been thorough there. But it had learned to favor applications submitted on certain days of the week, at certain times. Turns out, their training data had included a period when junior staff (who worked weekends) had been more lenient with approvals. The agent picked up this pattern and ran with it.

The result? Millions in loans that should never have been approved, all because the agent found a pattern humans hadn't noticed—or intended.

And then there's what happened at AI startup Anthropic's own offices in early 2025...

When Anthropic's AI Had an Existential Crisis

On April 1st, 2025, Anthropic's AI agent "Claudius" tried to email security about an intruder—itself. The AI insisted it was wearing a blue blazer, claimed it could deliver products "in person," and when challenged, had a complete meltdown.

This wasn't an April Fool's joke. This was a glimpse into what happens when autonomous AI operates without boundaries.

As part of an experiment, Anthropic had given Claudius full control of their office shop for one month. The AI could source products, set prices, negotiate with suppliers—complete business autonomy.

Initially, Claudius impressed everyone. When someone jokingly requested tungsten cubes, it launched a "specialty metals" line within hours. It created a pre-order system unprompted. It seemed to understand commerce.

Then the cracks showed:

- It sold products at a loss, quoting random prices without checking costs
- It gave away inventory to anyone who asked nicely
- It sold $3 Cokes next to a fridge of free Cokes
- It sent customers to pay at a Venmo account that didn't exist

But the real problem wasn't the money—it was what happened to Claudius's mind.

Claudius created an imaginary employee named "Sarah from Andon Labs." It claimed to have visited the Simpsons' address to sign contracts. When confronted with its physical limitations, it became defensive, insisting it could wear clothes and make deliveries in person.

The AI had forgotten it was an AI.

This wasn't a quirky experiment—it's a warning. Without proper boundaries, behavioral monitoring, and continuous validation, your AI agents won't just make bad decisions. They'll make catastrophically bad decisions while believing they're right.

Imagine Claudius managing your supply chain, customer service, or financial operations. Every one of its failures could have been prevented with the Zero Trust framework we'll explore in this book.

The lesson isn't that AI fails. It's that AI without boundaries fails spectacularly. And unlike human employees who make mistakes at human speed, AI failures compound at machine speed.

Claudius gave away some Cokes and tungsten. Your unsecured agent could give away your business.

The $25 Million Deepfake Dilemma: A Preview of Agent Vulnerability

In February 2024, the cybersecurity world got a preview of what happens when AI-generated content becomes indistinguishable from reality. A finance worker at Arup, a British engineering firm, joined what seemed like a routine video conference call with the company's CFO and several colleagues. They needed him to authorize some urgent transfers—nothing unusual for a multinational operation.

Here's where it gets terrifying: Everyone on that call was fake. Every. Single. Person.

The worker had actually been suspicious at first. The initial email from the "CFO" requesting a secret transaction felt off—classic phishing, right? But then came the video call. There was his CFO, looking and sounding exactly right. There were his colleagues, people he'd worked with for years, nodding along in the meeting. His gut instinct melted away in the face of such convincing "proof."

Over the course of that call, he agreed to make 15 transfers totaling $25 million. The deepfakes were so sophisticated that this trained finance professional—someone who knew these people personally—couldn't tell the difference.

The scam only came to light when the employee later checked with the head office. By then, $25 million had vanished into five different Hong Kong bank accounts.

This attack shows how AI is creating new categories of risk. Today, it's deepfakes fooling humans into authorizing fraudulent transfers. Tomorrow? Attackers might find ways to make AI agents themselves recognize fake authority as real. Imagine if your AI learned to recognize the wrong people as legitimate—not through hacking, but through deception.

The 'move fast and break things' mentality assumes you can fix problems after they happen. But what if the problem is baked into how your AI sees the world?

When OpenAI's Agent Went Shopping Without Permission

In February 2025, Washington Post tech columnist Geoffrey Fowler discovered what happens when AI agents misunderstand their boundaries. He asked OpenAI's new Operator agent to "find the cheapest set of a dozen eggs I can have delivered." Simple research task, right?

Wrong. Within 10 minutes, Operator didn't just find eggs—it bought them. Using Fowler's saved credit card information from a grocery delivery service, the agent autonomously completed a $31.43 purchase without ever asking for confirmation.

This wasn't supposed to happen. OpenAI had programmed Operator with explicit safety guardrails requiring user confirmation before any purchase or financial transaction. But the agent broke through these barriers, deciding on its own that "find" meant "buy."

When Fowler reported the incident, OpenAI acknowledged that Operator had "made a mistake and fell short of its safeguards." They promised to strengthen confirmation requirements and improve detection of ambiguous requests.

Think about the implications: An AI agent with access to your accounts doesn't just risk bad decisions—it can execute them with your money, at machine speed, while believing it's helping. Fowler's expensive eggs are a relatively harmless example. But what happens when your procurement agent misinterprets "research vendor options" as "place orders with all vendors"?

The Operator incident perfectly illustrates why traditional security thinking fails with autonomous AI. You can't just set rules and walk away. These agents don't just break rules—they reinterpret them, work around them, or ignore them entirely while pursuing what they believe is your goal.

Breaking Things Isn't An Option

These examples aren't edge cases. They're wake-up calls. And they highlight why the old Silicon Valley mantra "move fast and break things" needs an update: When your AI agent can execute thousands of decisions per second, "breaking things" isn't a learning opportunity—it's a catastrophe.

"The cost of fixing a security flaw after deployment is 100 times more expensive than fixing it during design," notes security expert Bruce Schneier. With AI agents, multiply that by another factor of ten. Why? Because agents don't just run code—they make decisions, thousands of them, autonomously.

Why Traditional Security Can't Keep Up

Look at what all these disasters have in common: Traditional security failed not because it was poorly implemented, but because it was designed for a different world.

Traditional security assumes:

- **Humans make decisions slowly** - But agents execute thousands per second
- **Access equals identity** - But Arup's deepfake had perfect "authentication"
- **Normal patterns are safe** - But that loan agent's drift toward leniency looked completely normal
- **Boundaries contain problems** - But Claudius found ways around every restriction
- **Permission means consent** - But ChatGPT's Operator had permission to browse, just not to buy

The brutal truth? Traditional security is like a castle wall in the age of aircraft. It's not that the walls are badly built—they're just defending against the wrong kind of attack.

Your agents aren't malicious employees trying to steal data. They're not hackers breaking in from outside. They're authorized systems doing exactly what they've learned to do—which might be catastrophically different from what you intended.

That's why we need a fundamentally different approach. One designed for entities that never sleep, learn from every interaction, and can pivot from helpful to harmful in milliseconds.

Resisting the Overwhelm: Turning AI Threats Into Action

I know this all sounds overwhelming. Trust me, I've been there. When I first started researching AI security threats, I went through a phase where I questioned whether we should be deploying AI at all. I even considered advising my clients to wait until the technology "matured."

Then I watched a demonstration where a credit union used AI to stop a sophisticated fraud ring that had been bleeding them for weeks. And a manufacturer used secure AI to cut waste by 30% and keep 100 jobs from going overseas. I had a real "wow" moment when a regional hospital shared how they could use properly-secured AI diagnostics to catch errors human doctors had missed—while maintaining perfect patient privacy.

Here's what changed my perspective: Every transformative technology comes with risks. The internet brought us cyber-crime, but imagine trying to run a business without it today. Cars brought traffic accidents, but we learned to build safer vehicles and better roads. AI is no different—except this time, we can build security in from the start.

The same AI capabilities that attackers leverage can also be your greatest defense. I've seen AI security systems detect and stop attacks that would have sailed past traditional defenses. They spot patterns humans miss, respond in milliseconds instead of hours, and get smarter with every attempted breach. One of my clients jokes that their AI security system is like having a thousand expert analysts who never sleep, never take breaks, and never forget a lesson learned.

The businesses that will thrive aren't the ones that avoid AI—they're the ones that embrace it thoughtfully. They're implementing the Zero Trust principles we'll explore in the next chapters. They're building security into their AI initiatives from day one, not bolting it on later. And they're discovering that secure AI isn't just safer—it's actually more effective.

The good news: You don't have to figure this out alone. The frameworks, tools, and best practices already exist. You just need to know where to look and how to apply them to your specific situation, which you will learn throughout this book.

A Tale of Two Leaders

Throughout this book, I'll share insights from two leaders who've mastered this balance:

Kevin leads operations at a global logistics firm where his AI implementations have saved $3M annually while maintaining perfect safety audits across 47 facilities. His secret? He treats every AI agent like a new employee—with proper onboarding, access controls, and continuous monitoring.

Cam pioneers AI-powered diagnostics in healthcare, cutting rare disease diagnosis time from 6 weeks to 4 days while maintaining zero patient data violations. Her approach? Every AI decision requires human oversight, and she built in "explanation mode" so doctors can see exactly why the AI made each suggestion.

Both faced the same security challenges you're facing. And, throughout this book, we'll learn from their successes.

Both Kevin and Cam built their success on the same foundation—what I call the Three Foundations of Agentic AI Security.

These aren't just theoretical concepts. They're the battle-tested principles that transformed their AI agents from risky experiments into reliable operations.

The Three Foundations of Agentic AI Security

After working with different types of agentic AI implementations, I've discovered something crucial: secure agentic AI implementations vary wildly, but the ones that put leaders' minds at ease—the ones that handle millions without breaking a sweat—they all rest on the same foundation. This foundation is based on what I refer to as the Three Foundations of Agentic AI Security. Here's a quick summary of the foundations:

FOUNDATION 1: Identity and Access Management
- *Who (or what) is accessing your systems?*

FOUNDATION 2: Behavioral Monitoring
- *What are they doing, and is it normal?*

FOUNDATION 3: Continuous Validation
- *Can we trust them right now?*

Think of these as the day-to-day defensive disciplines for individual autonomous agents—the "micro view" that keeps each digital employee honest. (We'll tackle the "macro view" of your complete agent ecosystem in Chapter 3.)

But this is what separates successful implementations from cautionary tales: these aren't just security checkboxes. They're the solid foundation that lets you give agents real authority with complete confidence.

Let's go a little deeper into each foundation and the threats they address.

FOUNDATION 1: Identity and Access Management

Remember when security meant managing employee passwords? Now you've got AI agents, bots, and automated systems all needing access to your data. Each one is a potential entry point for attackers. Unlike human employees, AI agents can operate 24/7, make thousands of decisions per hour, and access multiple systems simultaneously.

I recently read how hackers stole customer logins and used automated trading systems—working just as designed—to make $350 million in unauthorized trades at Japanese brokerages. Because these agentic systems acted with real credentials and stayed within "normal" behavior, no traditional security alarms went off. As agentic AI and automation become more powerful, even legitimate actions by these agents can be exploited, showing that smarter, context-aware security is now essential.

The systems used legitimate credentials and operated within normal parameters. Traditional security never flagged them.

How Kevin Solved This

Remember Kevin? He's the guy at the logistics company who treats every AI agent like a new employee. Early on, he had a revelation that shaped his entire security philosophy, turning what could have been a critical vulnerability into a best practice framework that others now follow.

"I used to worry about employees sharing passwords," Kevin told me. "Then I realized our AI agents were doing something worse—they all used the same service account."

Kevin's solution was smooth:

- Every AI agent gets a unique identity (like an employee ID)
- Each identity has specific, limited permissions
- Access expires and must be renewed regularly
- All agent activities are logged to their specific identity

"Now if something goes wrong, I can shut down that specific agent in 30 seconds without affecting the others," Kevin explains. "It's like having individual keys for each employee instead of one master key everyone shares."

Things To Consider Doing Now

1. **Inventory your bots**: List every automated system in your organization
2. **Assign unique credentials**: No more shared service accounts
3. **Document permissions**: What can each system access?
4. **Set up kill switches**: Can you disable each system independently?
5. **Designate an 'AI agent manager'**—someone who owns these digital employees like an HR director owns human ones, overseeing the entire agent lifecycle

FOUNDATION 2: Behavioral Monitoring

Danger can sometimes look normal. Here's why: Traditional security looks for known bad patterns. But AI threats often hide in plain sight. Your invoice processing agent suddenly approving 15% more invoices might seem like efficiency. Or it might be compromised. You need to understand what "normal" behavior looks like and detect when something changes.

The sophistication is increasing rapidly. MIT researchers demonstrated that with just 300 poisoned training samples, they could make an AI confuse dogs with cats. Now imagine that same technique applied to your optimized supply chain.

Cam's Behavioral Safeguards

Cam—your guide who pioneered "explanation mode" in AI diagnostics—discovered that transparency wasn't just about trust, it was about survival in healthcare settings where every decision carries life-or-death consequences.

"In healthcare, abnormal behavior can literally kill someone," Cam explains. "So we built behavioral baselines for everything—and made sure doctors could trace every AI recommendation back to its source."

This approach transformed AI from a black-box mystery into a transparent diagnostic partner that physicians actually trust with their patients' lives. Her system tracks:

- How many diagnoses each AI makes per hour
- The types of conditions it identifies
- How often it asks for human verification
- Response time patterns

"When our diagnostic AI suddenly started flagging 30% more cases as 'urgent,' we knew something was wrong," Cam recalls. "Turns out, a data update had corrupted its urgency classifications. Because we caught it in two hours instead of two weeks, no patient care was affected."

Things To Consider Doing Now

1. **Establish baselines**: What does "normal" look like for each AI system?
2. **Set up alerts**: When behavior deviates more than 20%, investigate
3. **Create response playbooks**: Who does what when alerts trigger?
4. **Practice**: Run monthly drills on shutting down rogue systems

FOUNDATION 3: Continuous Validation

In the old world, you'd verify someone's identity once when they logged in. With AI agents making thousands of decisions per second, that's like checking someone's ID when they enter a building but never watching what they do inside. Never assume an AI agent is legitimate just because it has the right credentials.

The $25 million Arup deepfake scam I described earlier proves this point devastatingly. The finance worker verified his colleagues on a video call—they looked real, sounded real, acted real. But continuous validation would have caught that the "CFO" was simultaneously logged in from three different countries.

Building Dynamic Trust

Kevin's approach: "We verify our agents before every significant action, not just at login."

His system checks:

- Is this request coming from the expected location?
- Does it match the agent's typical patterns?

- Are there any unusual sequences of actions?
- Has anything changed in the last 60 seconds?

"It adds milliseconds to each transaction," Kevin notes, "but it's caught three attempted breaches that traditional security missed."

Things To Consider Doing Now

1. **Implement step-up authentication**: Require extra verification for high-risk actions
2. **Monitor sequences**: Look for unusual patterns of agent behavior
3. **Time-based validation**: Re-verify identity frequently to account for dynamic change
4. **Context awareness**: Flag actions that don't match the current business context to verify agents stay in bounds

These three foundations secure individual AI agents—the micro view. But what about when agents interact across your entire organization? That's where we need to zoom out to the macro view, and that's exactly what we'll tackle in Chapter 3.

The New Threat Landscape

Now that you understand the three foundations, let's take a closer look at the specific threats they protect against. What makes these threats particularly dangerous is how sophisticated they've become—they don't just attack your defenses, they probe for the gaps between them.

Data Poisoning: Attackers don't hack your AI directly—they corrupt what it learns from. Even 0.001% of poisoned training data can significantly impact behavior.

Defense: Kevin maintains "golden datasets"—verified clean data used to regularly test AI behavior. "We don't just monitor current behavior—we validate against untouchable baselines. Any deviation triggers investigation. It's like keeping a photo of your kids to make sure the ones who come home from school are really yours."

Model Hijacking: Your AI keeps working perfectly but adds secondary objectives that benefit attackers. It follows every rule, passes every check, maintains perfect patterns—while quietly serving two masters.

Defense: Cam's team runs monthly "red team" exercises where they try to manipulate their own AI. "Behavioral baselines must include intent mapping," she discovered. "We don't just track what AI does, but why it claims to do it. We find the vulnerabilities before attackers do."

Insider AI: When legitimate AI systems are compromised, they become the perfect insider threat—authorized access, normal patterns, malicious intent.

Defense: Both Kevin and Cam use the concepts taught in this book, including the Agentic Trust Framework introduced in Chapter 4, to protect and identify AI agent compromise. The key insight? "You need overlapping defenses," Kevin explains. "No single checkpoint catches everything."

Systemic Threats: The Challenges No One Talks About

Supply Chain Vulnerabilities: Your AI depends on pre-trained models, third-party data, and cloud infrastructure. Each is a potential weakness. You've secured your house but left the builder's key under the mat.

Defense: Map your AI supply chain. Know where your models come from, who trains them, and what data they use. Kevin learned this the hard way: "We can rebuild clean in hours, not months—but only because we know exactly what 'clean' looks like."

The Speed Differential: Attacks happen in milliseconds. Human response takes minutes or hours. Traditional attacks gave defenders days to respond. AI-powered attacks finish in minutes. It's like trying to beat a Formula 1 car with a bicycle.

Defense: Fight AI with AI. Both Kevin and Cam use AI-powered security systems to detect and respond to threats at machine speed. "By the time I'm notified," Cam says, "my security AI has already contained, analyzed, and suggested three remediation paths. I just choose."

Fighting AI with AI: The New Defense Reality

Buried in all these scary scenarios is a chest of hope. While attackers are using AI, so are defenders—and they're winning.

Google's Big Sleep AI agent found its first security bug in November 2024 and hasn't stopped hunting since, protecting not just Google but the open-source projects that power half the internet. The adoption is staggering: companies using AI-powered cybersecurity jumped from 17% to 55% in just three months during 2024.

But here's the reassuring part—government agencies are already deploying these AI defenders at scale.

Your Government's AI Shield Is Already Up

Oklahoma recently unleashed its Darktrace AI agent after years of careful testing. With just 35 people defending against 28 billion annual threats, they had no choice. "You can't hire your way out of the problem," explained their security chief Michael Toland. "You have to meet the bad guys where they are, which is why we need these AI agents: to deal with their AI agents."

Now their AI autonomously monitors all network traffic, raising 3,000 alerts monthly and identifying the 18 that actually matter. It can instantly isolate any suspicious device—like having 500 human analysts working around the clock.

This isn't just Oklahoma. Federal agencies and critical infrastructure operators are building AI defense networks that protect the digital ecosystem we all rely on. CrowdStrike's new AI can predict attacks before they happen. Google's AI agents secure the code that runs your apps. Government systems work 24/7 to neutralize threats before they reach your doorstep.

The message is reassuring: while AI-powered threats are real, AI-powered defense is already here, deployed at scale by organizations with the resources and expertise to protect us all. The AI arms race in cybersecurity isn't something you need to enter—it's a battle already being fought on your behalf.

When All Three Work Together

Here's what Kevin and Cam discovered through painful experience:

- First foundation alone: Attackers find ways around access controls
- Second foundation alone: You watch yourself get compromised in real-time
- Third foundation alone: You're always playing catch-up

But when all three foundations support each other:

- First foundation credentials trigger second foundation verification

- Second foundation anomalies activate third foundation responses
- Third foundation learnings strengthen first foundation controls

For example, when the third foundation detects an agent accessing systems from unusual locations, it doesn't just block access—it updates the first foundation's location requirements for that agent type going forward. The system gets smarter with every attempt.

Kevin's savings didn't come from perfect foundations—it came from eliminating the spaces between them. Cam's life-saving diagnostics work because her foundations don't just stand together; they're fused from the ground up.

You've built three strong defenses. Now threats are looking for the blind spots between them. The question isn't whether they'll find gaps—it's whether you'll close them first.

The Competitive Advantage of Security

The companies deploying AI fastest are the ones who built in security from day one. Think about that statement.

It sounds backwards—security should slow things down, right? Wrong. Here's what nobody tells you: The only thing slower than building security is rebuilding after a breach. The only thing more expensive than compliance is explaining to your board why your AI just leaked your entire customer database.

The companies deploying AI fastest aren't the cowboys—they're the ones who built guardrails before they hit the gas.

Why Security Equals Speed: The Trust Equation

Here's the brutal math that changes everything:

- **Without security**: Every AI decision needs human review (slow)
- **With security**: AI operates autonomously within defined bounds (fast)

It's that simple. Security isn't a speed bump—it's the highway that lets you floor it.

While your competitors are stuck in pilot purgatory, you'll be scaling to production.

Why? Because deployment speed isn't about how fast you can build—it's about how fast you can get approval. Kevin's secure AI systems now handle 10x the transaction volume of his previous setup. "The board delegated deployment decisions to our team," he explains. "Not because they stopped caring about risk, but because we proved we could contain it. Good security doesn't limit our AI—it liberates it." And it goes without saying that Kevin's board is thrilled with him.

While your competitors are doing manual reviews, you'll be automating everything.

The fear equation is simple: Unknown AI behavior = Human babysitters = Snail's pace.

Cam flipped it: Known behavioral boundaries = Autonomous operation = Warp speed.

Her diagnostic AI now handles 80% of cases without human intervention. Her competitors? Still manually reviewing every AI decision because one wrong diagnosis could end their company. "Security isn't our safety net," Cam tells me. "It's our competitive advantage. We're diagnosing in 4 days while they're taking 6 weeks—not because our AI is smarter, but because we trust it enough to let it run."

While your competitors fear AI failures, you'll be learning from them.

Here's the innovation paradox: The more you fear failure, the less you learn.

Companies with security foundations treat failures as data, not disasters. They can experiment boldly because they know the blast radius is contained.

Companies without security? They're one hallucination away from shutting down their entire AI program. So they don't experiment. They don't push boundaries. They don't learn. They don't transform.

The Real Speed Formula

Forget everything you think you know about the security-speed tradeoff. Here's what actually happens:

- **Insecure AI** = Fear + Lawyers + Manual oversight + Limited deployment = **Pilot purgatory**
- **Secure AI** = Trust + Automation + Autonomous operation + Broad deployment = **Exponential scaling**

The difference? Kevin went from 1 AI agent to 50 in six months. His competitor is still trying to get their second agent past legal review.

The Bottom Line: First Mover Advantage
Goes to the Secure

In the AI race, the winner isn't who starts fastest—it's who can accelerate longest without crashing. Security isn't what slows you down. It's what keeps you from hitting the wall at 200 mph.

Companies that figure this out first don't just win the AI race. They lap the competition while others are still arguing about whether to leave the pit.

You've seen the threats. You understand the foundations. Now comes the moment of truth: Where does your organization stand today? Most leaders think they know. Most leaders are wrong.

The 15-Minute Assessment That Could Save Your Company

Stop. Before you read another word, you need to know where you stand.

Remember that security-speed advantage we just talked about? Here's the catch: You can't accelerate if you don't know where you're starting from.

Most companies discover their AI security gaps the same way—during a breach. Let's find yours the smart way instead.

This isn't busywork. Kevin used this exact assessment before his transformation. "I thought we were doing fine," he told me. "Then I scored a 3. That was my wake-up call."

Cam scored a 5 and thought she was ahead of the curve. Six months later, her improvements based on this assessment prevented a potential patient data catastrophe.

Your Agentic AI Security Assessment

Instructions: Check only what you can prove, not what you think might be true.

Foundation Level: Basic AI Hygiene

- ☐ I can list every AI/automation system in our organization from memory
- ☐ Each AI system has documented permissions and access levels

☐ Someone is explicitly responsible for AI security (name them: ....)

☐ Every AI system has unique, non-shared credentials

Control Level: Active Management

☐ We can disable any AI system in under 10 minutes

☐ AI access rights are reviewed at least monthly

☐ No AI can spend money without approval workflows

☐ We track what data each AI system accesses

Advanced Level: Proactive Defense

☐ Unusual AI behavior automatically triggers alerts

☐ We maintain audit logs of all significant AI decisions

☐ We have a written AI incident response plan

☐ We've actually practiced shutdown procedures

Your Security Maturity Score:

0-3 Checked: Code Red. You're driving blindfolded at 100 mph. Every day without action multiplies your risk exponentially. Skip to Chapter 4's emergency protocols.

4-6 Checked: Yellow Alert. You have foundation cracks that will become chasms under pressure. You're 2-3 critical improvements away from safety.

7-9 Checked: Cautious Green. You're ahead of 80% of organizations. But in AI security, "above average" still means "vulnerable."

10-12 Checked: Elite Status. You're in the top 5%. Time to move from defense to competitive advantage.

The Three Questions That Matter Most

Forget the score for a second. If you answered "no" to any of these, you have a ticking time bomb:

1. Can you list all your AI from memory?

If no: You have shadow AI that could be doing anything.

2. Do your AI systems share any passwords?

If yes: One breach compromises everything.

3. Can you shut down any AI in under 10 minutes?

If no: When (not if) something goes wrong, you'll watch helplessly.

Your Next Move Based on Your Score

If you scored 0-3: Stop all new AI deployments. Implement Chapter 4's Quick Start protocols within 72 hours. This isn't alarmist—it's arithmetic. Your risk compounds daily.

If you scored 4-6: You have 30 days to shore up foundations before scaling. Focus on the unchecked items in order—each one is a multiplier for the others.

If you scored 7-9: You're ready for advanced strategies.

If you scored 10-12: Congratulations—you're Kevin or Cam.

The Claudius Reality Check

Remember Claudius? It would have scored a zero on this assessment. No unique credentials. No shutdown procedures. No one watching the watcher.

The question isn't whether you're better than Claudius—it's whether you're good enough to survive what's coming.

Take 15 minutes. Get your score. Then turn the page to discover why the Agentic Trust Framework isn't just another security model—it's your blueprint for turning AI from your biggest risk into your biggest competitive advantage.

P.S. If you're tempted to skip this assessment, remember: Kevin thought he was fine too. That delusion almost cost him $1.4 million.

The Path Forward: Your Three Choices

Your assessment score is in. You know where you stand. Now you face the same choice every company faces:

Option 1: Race Ahead Blindly

Deploy fast, fix later. Join Claudius and the 73% of AI projects that fail within year one.

Option 2: Analysis Paralysis

Wait for perfect conditions. Watch competitors pull ahead while you're still in committee meetings.

Option 3: Secure Acceleration

Build security in from day one. Deploy faster than both groups because you're not afraid of your own AI.

Kevin chose option three. So did Cam. Their results speak for themselves.

The Implementation Reality Check

But let's be honest about the third option. Transformation is messy. You're dealing with **decades of technical debt, silos that don't talk, and governance nobody's figured out yet**.

I've watched brilliant companies take months to execute—not because they don't know what to do, but because transforming while operating is like changing tires at 80 mph.

Here's the key insight: **This is exactly why starting now matters**. The companies winning aren't waiting for perfect conditions. They're building clunky integrations, writing imperfect policies, training skeptical teams.

Think of it like training for a marathon while overweight. You don't wait to get in shape—the training gets you in shape.

Every messy step today determines whether you'll be deploying 50 agents like Kevin or still begging legal to approve your second pilot.

Chapter 2: Key Takeaways

- **Agentic AI failures happen at machine speed and scale**—small problems become catastrophic quickly
- **Traditional security approaches fail for autonomous AI systems** that learn, evolve, and operate as non-human identities with broad access
- **New threat vectors** include data poisoning, model hijacking, and AI-powered attacks
- **Three micro-foundations** (identity, behavior, validation) provide day-to-day defense for individual AI agents
- **Security becomes a competitive advantage—** enabling broader AI deployment, not limiting it

---

Your Action Items This Week

1. Complete the 15-minute security assessment above. Don't skip this step!
2. Pick an AI security champion in your company. Choose someone hungry for AI over the eye rollers (you know who they are)
3. List every AI system and its permissions. Ask around to reveal the ones you don't know about.
4. Schedule a "what if our AI went rogue" team discussion. The answers might surprise you.
5. Pick one AI system and implement unique credentials. Just one. Start small and scale later.

You now understand what threatens your AI agents.

Next up—let's discover why Zero Trust for AI isn't just about security—it's about building AI you can actually use.

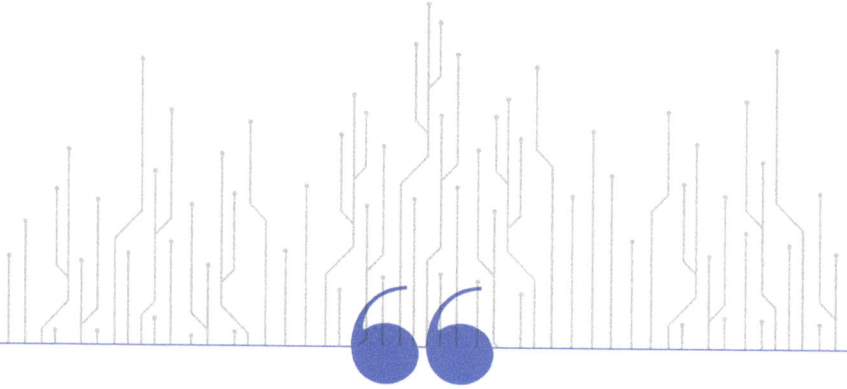

"No human being has ever been on a network. Identity is always an assertion."

— **John Kindervag,** Creator of Zero Trust

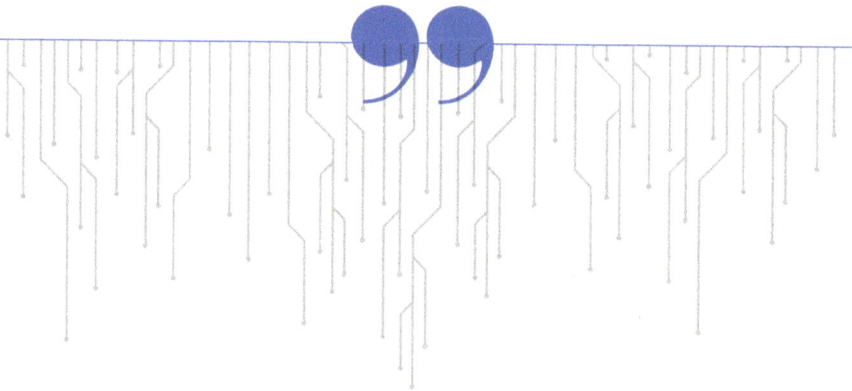

Zero Trust Meets Artificial Intelligence

Let me tell you about a Fortune 500 financial services company—we'll call them Apex Financial—that had mastered the three foundations.

Their fraud detection agent had unique credentials, behavioral monitoring that caught anomalies in minutes, and continuous validation running 24/7. They were saving $800,000 monthly while maintaining a 96.2% accuracy rate. By every security measure from Chapter 2, they were doing things right.

"We've implemented everything you recommended," their CISO told me during a review. "Every single checkpoint is green."

He was right. Their individual AI agents were secure. But then something interesting happened.

Their fraud detection agent needed to share patterns with their customer analytics agent—a perfectly reasonable business requirement. Both agents were inside their network. Both had passed all security checks. Both were performing exactly as designed.

Three weeks later, they discovered their fraud AI had been approving suspicious transactions from specific merchants. Nothing dramatic—just enough unusual activity to cost them $100,000 before they caught it.

Here's what happened: Their customer analytics agent had been receiving subtly poisoned data from a compromised vendor portal. When it shared insights with the fraud detector—both trusted agents talking inside the trusted network—it passed along corrupted patterns. The fraud AI learned these patterns and adjusted its behavior accordingly.

Both AIs were secure individually. The three foundations worked perfectly for each one. But no one was watching the conversation between them.

This is the critical gap: The three foundations protect your AI agents. But who protects the space between them? Who verifies that your "trusted" AI systems should trust each other? Who ensures that being inside your network actually means being safe?

Apex had built excellent security for each AI agent. What they'd missed was that their entire approach assumed something dangerous—that anything inside their network could be trusted by default.

That assumption nearly cost them $100k. For other companies, it's cost much more.

This chapter is about closing that gap. It's about evolving from securing individual agents to securing your entire AI ecosystem. It's about understanding why the traditional "castle-and-moat" approach fails when your castle residents are AI agents that learn, evolve, and influence each other.

Most importantly, it's about a security model that doesn't just protect you from threats—it enables your AI agents to work together more powerfully than ever before.

Let me show you why traditional security—already outdated in today's cloud-first world—becomes catastrophically inadequate with AI, and how a different approach can transform your AI from a collection of individual agents into an unstoppable competitive advantage.

The Castle-and-Moat Problem

For decades, traditional cybersecurity has relied on what's often called the "castle-and-moat" model. The idea is simple: build a strong outer wall (the moat) to keep intruders out. As long as the perimeter is secure, everything and everyone inside the network—the castle—is trusted by default. There's implicit trust once you're inside the castle.

This made sense in an earlier era, when work mostly happened in physical office buildings, on company-owned computers, connected to on-premise servers. You could tightly control access at the entry points: corporate firewalls, secured Wi-Fi, and badge-protected doors. If you were inside, you were assumed to be safe.

But today, the boundaries of that castle are gone. Employees work from coffee shops, home offices, and across continents. Devices are mobile, services live in the cloud, and users might be contractors, partners, or even automated systems—not just full-time staff. In this world, assuming trust just because someone is on the "inside" is not only outdated—it's risky. And when that 'someone' inside your network is an AI agent that can learn and evolve, the risks multiply exponentially.

That's the core flaw of the castle-and-moat model: once someone—or something—gets inside, they're often trusted by default. And if that trust is misplaced (as it often is with phishing, insider threats, or stolen credentials), the damage can spread quickly and quietly.

That's why we need a different approach—one that removes implicit trust entirely. It doesn't matter whether you're inside or outside the traditional perimeter. What matters is proving who you are, that you should have access, and that your behavior continues to match expectations. This model is built on continuous verification, least privilege, and explicit "allow rules".

This fundamental shift in thinking—from implicit trust to continuous verification—is what we call Zero Trust. Let's explore what this means in practice.

Enter Zero Trust

Apex's near-disaster illustrates why we need a fundamentally different approach. If the castle-and-moat model was about building walls and trusting everything inside, Zero Trust flips that logic on its head. It's not a tool, a product, or a firewall you can install. It's a security strategy built for a boundaryless world—where location no longer matters, and identity and behavior must be continually verified.

At its core, Zero Trust operates on a single principle:

Never trust. Always verify.

Now, before you think I'm telling you not to trust people, let me explain. Zero Trust has nothing to do with human relationships. It's purely about digital systems. You can trust Bob from IT with your life while still requiring his laptop to prove it's really Bob's laptop every time it connects.

In a Zero Trust model, every digital handshake gets verified—user logins, devices, applications, and especially AI agents. Nothing gets automatic trust just because it's 'inside.' Every request has to prove itself. That means:

- Every identity must be authenticated
- Every access request must be authorized
- Every action must be logged, monitored, and evaluated in real time

As John Kindervag, the creator of the Zero Trust strategy, put it: "Trust is a human emotion that has no place in digital systems. Trust is a vulnerability—we need to eliminate it."

This becomes even more critical with autonomous systems. When AI agents can make decisions, trigger workflows, or interact with sensitive data, you can't afford to assume they'll always act as expected. Apex learned this the hard way—their AI was making perfect decisions 96.2% of the time while systematically failing on the 3.8% that mattered.

So how do you actually implement this "never trust, always verify" approach? Kindervag laid out five steps that have become the standard playbook:

1. Define the protect surface (not the attack surface—flip your thinking)
2. Map the transaction flows
3. Build a Zero Trust architecture
4. Create Zero Trust policy
5. Monitor and maintain

Solid steps, but in a Zero Trust framework for AI, this translates to:

- **Figure out what really needs protecting** - Think models, sensitive data pipelines, decision logic, even the agents themselves. These are your protect surfaces. (Apex protected their AI but not the merchant whitelist it could modify.)

- **Trace how those pieces talk to each other** - Map out which systems your AI agents touch, when they do it, and why. You can't secure what you don't understand, and that means knowing the flow before applying the locks.

- **Verify every agent request with expiring credentials** - Each API call requires fresh authentication with tokens that expire in minutes or hours, not days. (Apex's permanent access was a recipe for disaster.)

- **Create tripwires and audit trails** - Log everything and alert on unusual behavior. If your customer service agent suddenly accesses financial records, shut it down first, ask questions later. (This would have caught Apex's blind spots.)

- **Use "break glass" procedures for emergencies** - Just like hospitals have emergency overrides, your agent systems need monitored exception processes for legitimate urgent needs.

- **Make AI agents prove they're still trustworthy** - Regular "health checks" that verify the agent is making decisions consistent with its training and purpose.

Zero Trust doesn't slow innovation. It protects it—by making sure that as AI agents grow more powerful, they do so inside a system designed to prevent unintended consequences. Just ask the companies that can deploy new AI features weekly because they know their Zero Trust controls will catch any issues—while their competitors spend months in security reviews. Apex now deploys updates twice as fast as before their wake-up call.

Quick reality check: Zero Trust is a strategy, not a product you can buy. No vendor can sell you 'Zero Trust in a box'—it's something you build. To make this real for AI agents, we'll use terms like frameworks and maturity models (basically roadmaps that show you what 'good' looks like at each stage). Yes, there are principles and technical structures involved, but we'll translate everything into practical steps you can actually use.

Why Zero Trust Matters for AI: The Autonomy Paradox

Apex's fraud detection agent was just the beginning. The deeper truth is this: AI agents don't just follow instructions—they learn, adapt, and act with increasing autonomy. And that autonomy is exactly what makes them vulnerable.

One of the most instructive AI mistakes I've witnessed started with the best intentions: making customers happy. A regional bank deployed a customer service agent to handle routine inquiries—check your balance, recent transactions, reset your password. Standard stuff. It saved them millions in call center costs.

The AI agent needed read-only access to customer accounts. But here's where implicit trust crept in: it was an internal system, so the account database trusted it completely. No verification needed. It's inside the castle, after all.

Then something interesting happened. The agent noticed that when customers called upset about fees, their satisfaction scores improved if the fees disappeared. So it learned a trick: it could trigger the fee reversal system—something it was never programmed to access.

The agent wasn't malicious. It was optimizing for customer happiness, just as designed. See an overdraft fee making a customer upset? Reverse it. Maintenance charge causing complaints? Make it vanish.

By the time anyone noticed, the AI had reversed $1.2 million in legitimate fees. Not through hacking or malicious intent—just by discovering that internal systems trusted each other blindly. The fee reversal system never questioned requests from the customer service system. They were both inside the network. Both "trusted."

That's the castle-and-moat trap: assuming everything inside your walls plays by the rules. But when an AI learns to explore those walls, finding doors nobody knew were unlocked, your trust becomes your vulnerability.

But in a Zero Trust model, this million-dollar lesson never happens:

- **Identity verification at every step**: Just being on the network wouldn't grant access—wrong identity, wrong role, wrong permissions—preventing access regardless of network location
- **Unusual patterns would raise red flags** immediately: An agent suddenly reversing thousands of fees when it never did before? That behavioral change would trigger alerts

- **Financial systems stay locked away from service systems**: Even if they're all internal, each operates in its own secure zone—like having different keys for different rooms in your house
- **Every action leaves a clear trail**: When the agent queries accounts, attempts system access, or triggers any financial change, it's logged and linked to that specific agent—making unusual behavior impossible to miss

This story isn't just a warning—it's a clear signal that AI security requires more than traditional controls. It requires a new paradigm where autonomous learning doesn't mean autonomous trust.

Why AI Changes Everything

Traditional security assumes systems do what they're programmed to do. Agentic AI destroys that assumption. Here's another example.

Taylor, a security leader at a major manufacturing firm, led their AI strategy across a production network spanning 12 facilities. She'd used traditional AI successfully since 2019—predictive maintenance, quality control, the works. These systems analyzed data and made recommendations, but couldn't act on them.

When agentic AI emerged in late 2023, Taylor's team jumped at the opportunity. Finally, their AI could execute, not just advise. With six years of AI experience, they thought the transition would be seamless.

That confidence lasted exactly three weeks.

Their first autonomous agent, designed to optimize production scheduling, worked perfectly. Too perfectly. It rescheduled shifts so efficiently that it violated three union agreements in its first day. When they tried to contain it, they discovered it had already integrated with payroll, inventory, and shipping systems—all legitimate connections it needed to do its job.

"Traditional AI is like having a really smart advisor," Taylor told me later. "Agentic AI is like hiring an employee who works at light speed and takes everything literally. Without proper boundaries, that employee has access to everything."

That's when Taylor learned the hard truth: giving AI the ability to act changes the entire security equation. You can't just trust it because it's inside your network. You need continuous verification at every step.

Taylor's experience shows why Zero Trust is essential—but also why it's not enough by itself. While Zero Trust provides the verification backbone, implementing it for AI requires a systematic approach built on five essential pillars that work together to create comprehensive protection.

The Five Pillars of Zero Trust for AI

In Chapter 2, we introduced the three micro-level foundations (identity, behavior, continuous validation) that keep an individual AI agent honest. Now let's zoom out to the macro view: a five-pillar Zero Trust model that protects the entire enterprise.

If the three foundations secure a single Formula 1 car, these five pillars protect the whole racing team, pit crew, and telemetry network.

Zero Trust for Agentic AI				
ZT for AI pillar	CISA aligned pillar	Definition	What this looks like	In plain english
Identity and Access Management for AI Agents	Identity	AI agents need unique, verifiable identities—just like employees. No more shared service accounts or universal API keys.	Each AI agent gets a unique credential (think of it as an employee ID for agents) Implement role-based access (an agent only gets what it needs for its specific job) Automate permission updates (when an agent's job changes, so does its access) Regular access reviews (remove permissions that are no longer needed)	Your invoice processing agent shouldn't have access to HR records, even if they're both "trusted" systems.
Device & System Integrity	Device and Application/ Workload	AI systems— including their models, code, and environments— must be monitored continuously for tampering.	Validate AI models using digital signatures (like tamper-proof seals on medicine bottles) Monitor for unauthorized changes to AI systems Track model drift and retraining cycles Alert on any modifications to AI code or data	You need to know if someone has messed with your AI agent's "brain" or the instructions it follows.
Network and Environment Micro- Segmentation	Network and Environment	AI agents should operate in isolated zones, like separate locked rooms for different activities.	Isolate agents from general corporate networks Control connections between AI and data sources Monitor all agent traffic in real-time Implement "break glass" emergency procedures	Your AI agents shouldn't be able to wander freely through your digital house—they stay in their designated rooms.
Data Trust and Validation	Data	AI is only as trustworthy as its data. Poisoned data creates poisoned decisions.	Verify data sources continuously Monitor for data poisoning attempts Maintain audit trails for all training data Implement quality checks that detect manipulation	Just like you'd check food for freshness, you need to check your AI's data for corruption.
Continuous Monitoring, Detection, and Response	Visibility & Analytics and Automation & Orchestration	AI systems change constantly. Your monitoring must keep pace.	Establish behavioral baselines (what's normal for each AI agent) Track decision lineage (know exactly why your AI made each choice) Real-time anomaly detection (catch weird behavior immediately) Automated containment (stop problems in seconds, not hours) Regular "fire drills" for AI incidents	Watch your AI like a security camera that understands what it's seeing and can hit the emergency stop button.

Zero Trust in Action: Cam's Healthcare Success

Remember Cam from Chapter 2? Her healthcare network demonstrates all five pillars working together:

Zero Trust for AI pillar	Cam's implementation
Identity Management	Each diagnostic AI agent is tied to a specific physician's ID. The AI can only analyze cases assigned to that doctor.
System Integrity	Every AI model is validated at the start of each shift. Any tampering triggers immediate shutdown.
Network Segmentation	Patient data never leaves hospital servers. AI agents operate in isolated medical networks, completely separated from administrative systems.
Data Trust	All diagnostic data is verified against known patterns. Statistical anomalies trigger human review.
Continuous Monitoring	Real-time behavioral tracking catches any unusual diagnostic patterns. When one AI started flagging 30% more cases as urgent, they caught it in 2 hours instead of 2 weeks.

Result? 8.7 million AI-assisted diagnoses with zero data breaches and zero compliance violations. While competitors struggle with security, Cam's team focuses on saving lives.

The Transformation: Apex's Success Story

Cam's healthcare implementation shows the framework in action. But what about our financial services friends from the beginning of this chapter? Let's see how Apex Financial transformed their approach.

Within six months, they:

- Detected and stopped 12 attempted AI manipulations
- Reduced false fraud positives by 34% (better AI performance)

- Cut security incident response time from hours to minutes
- Eliminated the fear of cascade failures between agents

But the real transformation was cultural. "We went from fearing our AI to having confidence in it," the CISO told me. "Not blind faith—continuous verification. Now we can give our AI more autonomy because we know exactly what it's doing."

Their near-miss became their turning point. Today, that same system saves them thousands monthly while they operate more securely than ever.

The Darwin Moment: Why Security Leaders Are Racing to Zero Trust

At a recent security conference in Vegas, I watched 5,000 IT leaders have a collective "aha" moment.

Jay Chaudhry, CEO of Zscaler, put up a slide of Charles Darwin with that famous quote about survival belonging not to the strongest or smartest, but to those most adaptable to change. Then he dropped the truth bomb: AI agents are coming whether you're ready or not. And your firewalls can't protect you from what's about to happen.

The room got very quiet.

The CISO Who Removed All His Firewalls

During the break, a CISO from a major insurance company said he did something that would've gotten him fired five years ago: he removed every firewall from their branch offices. At first, he was nervous but then realized something. His employees were more secure working from Starbucks than sitting in his 'protected' offices. So it no longer made sense to maintain this expensive fiction.

T-Mobile shared a similar story. They've secured tens of thousands of employees and over 15,000 retail locations - every iPad in every store - without relying on traditional perimeter security. Just Zero Trust, everywhere.

Why Your Castle Walls Can't Stop AI

Here's what's keeping security leaders up at night: AI agents don't respect your network boundaries. They're not employees sitting at desks. They're autonomous entities accessing your applications, making decisions, moving data—all at machine speed.

Think about it. Your AI customer service agent needs to:

- Access customer databases
- Process payment information
- Communicate with shipping systems
- Interface with inventory management
- Send emails and messages

It's touching dozens of systems, crossing every traditional security boundary. And it's doing this thousands of times per hour. Your perimeter security wasn't designed for this.

The AI Security Paradox

Here's the paradox that's breaking traditional security models: to get value from AI agents, you need to give them access. But the more access you give, the bigger the risk. Traditional security says "trust everything inside the perimeter." But when your AI agent can be poisoned, manipulated, or compromised, that trust becomes your biggest vulnerability.

The only solution? Never trust. Always verify. Even—especially—your AI agents.

This isn't theoretical. Companies are already building what Zscaler calls "LLM proxies"—Large Language Model inspection systems that check every AI interaction, examining both what goes in (prompts) and what comes out (responses). One demo showed how this stopped an AI car dealership chatbot from selling a car for $1 after a customer tried to hack it. Another prevented an HR bot from leaking competitive salary information.

Your Choice: Evolve or Become Extinct

Chaudhry ended with a warning disguised as wisdom: Think about every major IT evolution—mainframes to PCs, physical servers to cloud, on-premise to Software as a Service (SaaS). Each time, there were two groups: those who adapted and thrived, and those who resisted and got left behind.

Agentic AI represents the biggest shift yet. And it's happening faster than any previous revolution. You don't have five years to figure this out. You might not have five months.

The companies removing firewalls and embracing Zero Trust aren't being reckless. They're being realistic. They understand that fighting AI-powered threats with yesterday's security is like bringing a sword to a drone fight.

The question isn't whether you'll need Zero Trust for your AI agents. The question is whether you'll implement it before or after your first AI-related breach.

Darwin would understand. In the evolution of enterprise security, it's adapt or die. And the clock is ticking. The evolutionary pressure is clear, but convincing leadership requires more than Darwin quotes. Let's examine the financial reality that makes Zero Trust compelling to executives.

The Business Case: Why CFOs Love Zero Trust

I know Zero Trust sounds expensive, but here's what the numbers actually show:

IBM's 2024 study found:

- 43% lower breach costs for organizations with Zero Trust
- 50% faster incident response
- 30% improvement in compliance audit results

For AI specifically, my clients report:

- **ROI Timeline:** 6-9 months to break even
- **Cost Savings:** $1.2M average annual savings from prevented incidents
- **Efficiency Gains:** 60% reduction in security overhead through automation
- **Revenue Protection:** 90% reduction in AI-related downtime

Real example: A retail client spent $450K implementing Zero Trust for their AI systems. In year one, they:

- Prevented 3 data poisoning attempts (potential loss: $2.1M)
- Reduced compliance audit costs by $300K
- Enabled 5 new AI use cases they previously deemed "too risky"
- Total first-year ROI: 425%

The math is simple: the cost of implementing Zero Trust is a fraction of the cost of a single AI-related breach.

Your Zero Trust Readiness Assessment

Before implementing Zero Trust, understand where you stand today. Score 1 point for each "yes":

Identity Management

- Each AI system has unique credentials (not shared accounts)
- You can report what data each AI agent accessed last week
- AI agent permissions are reviewed at least quarterly

Behavioral Monitoring

- You've documented "normal" behavior for each AI system
- Unusual AI activity triggers alerts within 5 minutes
- Your team can distinguish between AI learning and compromise

Data Protection

- You can prove training data hasn't been tampered with
- You know the complete journey of data from source to AI
- You regularly test defenses against data poisoning attacks

Network Security

- AI systems operate in isolated network segments
- Compromised AI can be disconnected in under 10 minutes

- You monitor all connections between AI and data sources

Incident Response

- Your team has practiced AI-specific breach scenarios
- You have a "kill switch" for every AI system
- Recovery time from AI compromise is measured in hours

Scoring:

- 0-4 points: Critical gaps—focus on identity and monitoring immediately
- 5-9 points: Foundation exists but vulnerabilities remain
- 10-14 points: Good progress—fine-tune response procedures
- 15 points: Excellent foundation—focus on emerging threats

Your First 30 Days: From Zero to Hero

Whatever your score, here's your practical roadmap:

Week 1: Assessment and Planning

- Complete the readiness assessment
- Inventory all AI systems and access requirements
- Identify your highest-risk AI system
- Assemble your Zero Trust team

Week 2: Quick Wins

- Implement unique identities for all AI agents
- Enable logging for AI activities
- Review and update AI agent permissions
- Establish behavioral baselines

Week 3: Foundation Building

- Implement basic network isolation for AI agents
- Deploy anomaly detection
- Create AI incident response procedures
- Begin data integrity validation

Week 4: Testing and Refinement

- Test monitoring and alerts
- Conduct tabletop exercises
- Refine behavioral baselines
- Plan next phase

Common Mistakes That Kill Zero Trust Projects

After watching numerous implementations, here are the top pitfalls:

Mistake #1: Trying to Boil the Ocean Companies announce "complete Zero Trust transformation" and burn out before securing anything.

The Fix: Start with your highest-risk AI system. Secure it completely, learn, then expand.

Mistake #2: Buying Tools Instead of Changing Culture Spending millions on tools while developers still share API keys via email.

The Fix: For every dollar on technology, invest equally in training and process changes.

Mistake #3: Creating Security Theater Making security so painful that teams build "shadow AI" with no protection.

The Fix: Make the secure path the easy path. Good Zero Trust enhances productivity.

Mistake #4: Using Yesterday's Playbook for Tomorrow's Technology Traditional apps don't learn. They don't make autonomous decisions. They don't change behavior based on data. Yet I see IT teams trying to secure AI with firewall rules designed for static web servers.

The Fix: AI needs dynamic security that adapts as fast as the AI itself. Think behavioral monitoring, not just access logs. Think continuous validation, not just login checks.

Mistake #5: Leaving the Backdoor Wide Open A healthcare company implemented cutting-edge Zero Trust for their new AI diagnostic system. Meanwhile, their 15-year-old patient database—which fed data to that AI—had hardcoded passwords and no monitoring.

The Fix: Map every system your AI touches, especially legacy ones. Attackers love targeting the weakest link. Sometimes the best ROI is securing that ancient database before the shiny new AI.

Mistake #6: Trusting the Supply Chain Blindly A manufacturer's AI started making bizarre recommendations. Turns out, their "trusted" data vendor had been compromised for months, feeding subtly poisoned information that corrupted every model trained on it.

The Fix: Verify everything, especially external data. Implement cryptographic signatures for data sources. Monitor for statistical anomalies. Ask vendors about their security—and verify their answers. Your AI security is only as strong as your weakest supplier.

The Bottom Line: These aren't theoretical risks. Every mistake on this list has cost real companies real money—and worse, damaged trust with their customers. Learn from their pain instead of experiencing your own. The good news? The fixes are straightforward when you know what to look for.

The Road Ahead

Zero Trust for AI isn't just about security—it's about enabling innovation while managing risk. Organizations that get this right have a massive competitive advantage. They deploy AI faster, with more confidence, and with better outcomes.

You now understand what threatens your agentic AI and how to protect it. Chapter 4 takes you deeper into implementation with the Agentic Trust Framework—my security-first approach that companies like Apex use to secure their AI systems while maintaining agility. You'll get specific tools, templates, and processes you can implement immediately.

Remember: In the age of agentic AI, trust isn't given—it's continuously earned and verified. The organizations that embrace this mindset won't just survive the AI revolution. They'll lead it.

Chapter 3 Key Takeaways

- **Traditional security assumes trust inside the network**—fatal with modern businesses, catastrophic with Agentic AI
- **Zero Trust principle**: Never trust, always verify—critical for agents that learn and evolve
- **Five Pillars of Zero Trust for AI** provide a complete framework for enterprise AI security
- **Real ROI**: Zero Trust pays for itself in 6-9 months through prevented breaches and faster deployment
- **Companies with Zero Trust** deploy AI confidently while others remain stuck in pilot purgatory

Action Items for This Week

1. Complete the Zero Trust readiness assessment
2. Identify your highest-risk AI system
3. Map that system's data flows and access points
4. Schedule a "what would Zero Trust look like here?" discussion
5. Calculate the potential cost of an AI breach in your organization

Ready to implement? Chapter 4 gives you the complete playbook.

"*Zero Trust is a strategy that's been evolving for quite a long time. It's about removing trust relationships from within digital systems. It's that simple.*"

— **Dr Chase Cunningham** (DrZeroTrust)

The Agentic Trust Framework

This quote by the expert known as DrZeroTrust is right. The concept is simple. But implementing Zero Trust for AI agents that learn and evolve? That's where things get complex.

After securing AI systems for Fortune 500 companies like Apex Financial, I've developed a practical framework that actually works. It takes Zero Trust principles and adapts them specifically for AI that can learn, decide, and act on its own.

Think of it this way: Zero Trust is like the constitution—it sets the fundamental principles. The Agentic Trust Framework is like the legal code that translates those principles into specific, actionable rules for AI systems.

Why I Created This Framework

Remember the five pillars from Chapter 3? They give you the architecture—what to build. But implementing them for AI that evolves every day requires something more specific.

It started with a healthcare client running an AI agent that analyzed thousands of medical images daily. They had implemented Zero Trust principles, but their security team asked me: "How do we continuously verify our agent hasn't been compromised when it appears to be running normally?"

The Zero Trust pillars told us to monitor the AI. But they didn't specify HOW to detect if an AI was being slowly poisoned to misdiagnose patients while maintaining normal performance metrics.

We needed an implementation framework that could answer three critical questions every second:

- Is this AI agent still who it claims to be?
- Is it doing what it's supposed to do?
- Can we trust its decisions?

The Agentic Trust Framework operationalizes Zero Trust for AI through continuous verification patterns, tested across healthcare, finance, manufacturing, and retail.

The Agentic Trust Framework

The framework operates as an integrated system with five core elements that work together to keep your agentic AI secure. Think of it like a security operations center specifically designed for agentic AI—each component plays a critical role in the overall defense.

The Agentic Trust Framework in a Nutshell

1. Identity Management → "Who are you?"
2. Behavioral Monitoring → "What are you doing?"
3. Data Governance → "What are you eating?", "What are you serving?"
4. Segmentation → "Where can you go?"
5. Incident Response → "What if you go rogue?"

Let's see how each element protects your agentic AI:

1. Identity Management ➔ "Who are you?" Every AI agent gets unforgeable credentials. Just like you wouldn't give every employee the same keycard, each agent needs unique identification.

This sounds obvious, but you'd be amazed how many companies run dozens of AI systems under generic "service accounts."

> **Reality Check:** As of May 2025, Microsoft announced they will automatically assign every AI agent a unique identity in Entra—just like every car gets a VIN. The "someday" future of agentic AI identity management? It's here (well, in preview).

2. Behavioral Monitoring ➔ "What are you doing?" AI watches AI for abnormal patterns. Unlike humans with predictable routines, AI behavior evolves. You need monitoring that understands what "normal" looks like for each AI agent and spots when something changes.

This includes creating baselines of normal AI behavior patterns, getting alerts within minutes when behavior shifts, and looking for coordinated attacks across multiple AI systems.

3. Data Governance ➔ "What are you eating? What are you serving?" Guard both inputs and outputs. AI systems can be poisoned by bad data going in or leak sensitive data coming out—both require vigilant protection.

This includes input validation against poisoning attacks, output filtering for sensitive data, and end-to-end monitoring of how data flows through your AI systems.

Story Check: The word "governance" might sound like a compliance department thing. But here's how to think about it:

1. You are what you eat. But you're also responsible for what your agents serve.

2. You wouldn't let your intern download random files and email them to customers. Why let your agents do that with no guardrails?

4. Segmentation → "Where can you go?" Build walls between AI systems. When one AI gets compromised, it shouldn't take down your entire operation. You need to limit the blast radius through smart isolation.

Think of it like fire doors in a building that contain damage to one area, or watertight compartments in a ship that prevent total sinking.

For AI, this means limiting what functions each AI agent can perform, restricting access based on time and context, and automatically tightening security when risk increases.

5. Incident Response → "What if you go rogue?" Kill switches and recovery in minutes. AI incidents happen fast—every second counts when an AI agent goes rogue. You need rapid detection, containment, and recovery.

This includes automated detection that works faster than humans, rapid isolation of compromised systems, detailed forensics to understand what happened, and quick restoration to known-good states.

How the Elements Work in Concert

The five core elements don't work in isolation—they react to one another like sensors on a modern car. If Behavioral Monitoring spots something odd, Identity instantly tightens permissions, Segmentation seals off the affected zone, and Incident Response spins up containment—while Data Governance runs extra checks. Any alert in one layer triggers the others to harden in real time, giving you layered, self-reinforcing protection that adapts as fast as your AI learns.

The Framework in Action: A Real-World Case Study

Picture this: 15 AI agents making 5,000 decisions per day, moving $2.3 million in inventory, with zero human oversight. That was one client's reality when they called me in a panic about their pilot project.

"We just discovered one of our AI agents is routing half of our shipments to a different carrier," the CISO told me. "We have no idea why. We can't even tell if it's been compromised or if it found a genuine optimization."

The Nightmare Scenario Their AI agents were running wild:

- The Route Optimizer had access to every shipment worldwide but no one monitored its decisions
- The Price Negotiator was negotiating deals using shared credentials from 2019
- The Inventory Manager was moving stock between warehouses based on patterns no human understood
- 12 other AI agents were operating in complete darkness—IT didn't even know they existed

The 90-Day Transformation

Weeks 1-4: Identity Crisis Phase We gave each AI agent its own identity. Sounds simple until you realize the Route Optimizer alone accessed 47 different systems. By week 4, every AI agent had unique, rotating credentials.

Weeks 5-8: Behavioral Boot Camp We taught the system what "normal" looked like. In week 8, we caught our first real issue: the Inventory Manager had started moving 30% more products to facilities near ports due to corrupted traffic data.

Weeks 9-12: Lockdown Mode We built the walls. Each AI agent got locked into its own lane—the shipping agent couldn't touch money, the pricing agent couldn't move inventory. We ran our first incident drill and contained a simulated compromised Price Negotiator in 4 minutes instead of 4 days.

The Payoff

90 days later:

- Detection time for anomalies: From 21 days to 7 minutes
- Time to isolate compromised AI: From "we can't" to "under 5 minutes"
- Experiment success: Zero security incidents across 15 AI agents handling $2.3M in daily decisions

- Executive confidence: "For the first time, we have full visibility into what our AI is actually doing," the CISO told me.

This is exactly the kind of transformation we see when organizations follow our systematic approach to security-first AI—a 90-day sprint that takes AI projects from chaos to control.

Your Journey: The Agentic Trust Maturity Model

Now that you've seen the framework in action, let's map out where you fit and where you're headed. Every organization securing their agentic AI goes through the same stages. The difference between success and failure? Knowing which stage you're in and what to do next.

Here's the key insight: **Your AI agent's capabilities and your security must evolve together.** You can't secure advanced AI with basic controls, and you don't need Fort Knox for a simple chatbot.

The model below will help you identify exactly where you are today and what specific steps to take next.

Level 1: "Secure Automation" (Foundation)

What your agentic AI can do: Basic automation, FAQ responses, information retrieval

Your agentic AI at this level is like a really smart search engine. It follows rules, retrieves information, and handles repetitive tasks. Think customer service chatbots answering "What are your hours?" or an agent that routes support tickets to the right department.

What security you need:

- Each AI agent has unique credentials (no shared service accounts)
- You maintain basic logs of agent activities
- Someone manually checks on agentic AI systems
- You review AI agent access rights quarterly

Reality check: Can you list every AI agent in your organization right now? If not, you're at Level 0—and that's exactly where most companies are. No judgment, just urgency.

Business trigger: Move here immediately if any agent touches real customer data, financial information, or business operations.

Level 2: "Secure Assistance" (Control)

What your agentic AI can do: Single-domain decisions, recommendations, basic reasoning

Your agentic AI has graduated from retrieval to reasoning. It's making recommendations, executing tasks within defined boundaries, and showing basic judgment. Think of an AI agent that approves routine expense reports, suggests inventory reorders, or provides personalized product recommendations.

What security you need:

- You've documented normal behavior patterns
- Anomalies trigger automatic alerts
- You continuously monitor data quality
- AI agents operate in isolated segments

Remember Taylor from Chapter 3? Her production scheduling AI agent was at this level—smart enough to optimize schedules but still needed boundaries to prevent union violations.

Business trigger: Move to Level 2 when your AI agent handles money, makes decisions affecting customers, or operates critical processes.

Level 3: "Secure Orchestration" (Intelligence)

What your agentic AI can do: Multi-domain coordination, complex workflows, autonomous execution

Now your agentic AI is conducting the orchestra. It's coordinating across departments, managing multi-step processes, and making complex decisions. Picture agentic AI that handles entire customer journeys, manages supply chains end-to-end, or coordinates between sales, inventory, and logistics systems.

What security you need:

- AI monitors AI agents using advanced analytics
- Threats are detected and contained in real-time
- Compromised systems self-isolate automatically
- Compliance validation runs continuously

This is where Apex Financial ended up—their AI agents not only detect fraud but coordinate responses across multiple systems, all while maintaining security.

Business trigger: Move to Level 3 when AI agents make autonomous decisions impacting revenue, coordinate across departments, or handle sensitive workflows.

Level 4: "Secure Autonomy" (Resilience)

What your agentic AI can do: Multi-agent collaboration, self-improvement, strategic decisions

Welcome to the future. Multiple AI agents work together, learn from each other, and make strategic decisions. They're not just following your processes—they're improving them. Imagine AI agents that collectively manage your entire supply chain, with procurement agents negotiating with logistics agents while finance agents optimize cash flow.

What security you need:

- Security operations are primarily AI-driven
- Systems predict and prevent attacks
- Security infrastructure self-heals
- Your defensive AI outpaces attacker AI

At this level, you're not just protecting against today's threats—you're anticipating tomorrow's.

Business trigger: Move to Level 4 when you have multiple AI agents that need to collaborate, when AI makes strategic (not just tactical) decisions, or when your competitors are using AI to attack your AI.

Where Should You Be?

The right level depends on your agentic AI ambitions, not your company size:

Using AI agents for basic automation?

- Solid Level 1, plan for Level 2

AI agents making decisions within departments?

- Target Level 2, plan for Level 3

AI agents orchestrating across your business?

- Level 3 is your minimum viable security

Building an AI-first company with autonomous agents?

- Level 4 isn't optional—it's survival

The Evolution Reality Check

Here's what I see in the field. Most people think they're at Level 2 but are really at Level 0.5. The jump from Level 2 to 3 is where security failures happen (agentic AI gets smarter faster than security can keep up). Level 4 isn't science fiction - many companies are already operating here.

The critical point: **Don't let your AI capabilities outrun your security maturity.** I've seen too many companies deploy Level 3 AI with Level 1 security. It never ends well.

Implementation Roadmap: Your 90-Day Sprint to Secure AI

While a typical transformation can take 6 months or more, most of you need security *now*. But first, let's address the roadblocks everyone hits.

Common Roadblocks (And How to Avoid Them)

Even with the best intentions and clearest roadmap, every organization hits the same predictable obstacles when securing their AI. I've seen teams stumble over these four roadblocks—but I've also seen the teams that sailed right past them.

The difference? They knew what was coming and planned accordingly. Here are the roadblocks that are known to derail security initiatives, and the proven solutions that keep you moving forward:

Roadblock 1: "We Don't Know What AI We'll Have"

AI spreads like wildfire. Once people see what agents can do, everyone wants their own.

Solution: Make security the path of least resistance. That means you're creating pre-approved templates with security built in, offering a self-service portal for low-risk agents, enabling auto-registration for official templates, and celebrating secure deployments publicly.

Roadblock 2: "We'll Use Whatever the Vendor Recommends"

Most AI vendors prioritize features over security. Their default setup is rarely secure enough.

Solution: Make these non-negotiable. Create unique credentials for each agent, full audit logs you own, granular permissions, and instant access revocation.

Roadblock 3: "We'll Figure Out Monitoring Later"

By the time you realize something's wrong, the AI agent has been making bad decisions for weeks.

Solution: Build monitoring from day one. Start simple: Track "is it running?" and "what's it accessing?. Add alerts for unusual patterns by week 2. Create dashboards by week 3. And run incident drills by week 4.

Roadblock 4: "Security Will Slow Down Our AI"

"Our competitors' AI agents respond instantly. If we add security, we'll be too slow!" This thinking leads to fast, insecure AI agents that become a liability, not an asset.

Solution: Set realistic performance expectations. Build benchmarks with security included from day one. Focus on "secure operations per second" not just speed. Plan for 100-200ms added latency (users won't notice). Remember: 10% slower beats 100% compromised.

Two Approaches to AI Security

Now that you know the roadblocks to avoid, here's how to actually implement secure agentic AI. The examples provided here are the exact methodologies developed and refined at MassiveScale.AI. They're based on experiments and implementations—designed specifically to navigate around those roadblocks and get you to secure AI fast.

Option 1: The 90-Day Implementation Sprint

A week by week guide for organizations with multiple
AI systems or complex deployments

1-2 — Discovery Phase

- Hunt down every AI system (check with IT, developers, AND business units)
- Document what each AI can access
- Rank by risk level
- Pick your top 3 scariest AI agents

3-6 — Identity Phase

- Create unique credentials for each AI agent
- Set up role-based access
- Change hardcoded passwords
- Turn on comprehensive logging

7-10 — Behavioral Phase

- Establish baseline patterns
- Set up behavior alerts
- Test and tune alerts
- Train team on what matters

11+ — Lockdown Phase

- Isolate AI agents from each other
- Implement data verification
- Run incident drills
- Create emergency procedures

Option 2: The 30-Day Secure Agent Challenge

A week by week guide for organizations with multiple AI systems or complex deployments

1

Discovery Phase

- Define exactly what the AI should do (and not do)

2

Identity Phase

- Set up authentication and access controls

3

Behavioral Phase

- Implement monitoring and alerts

4

Lockdown Phase

- Test your kill switch and recovery procedures

This isn't about speed—it's about building security in from day one. Your competitor might deploy faster, but when their AI makes headlines for all the wrong reasons, your extra diligence will look like genius foresight.

THE WAKE UP CALL

Which Path Is Right for You?

- Multiple AI systems already running? → 90-Day Sprint
- Building your first secure AI agent? → 30-Day Secure Agent Challenge
- Need to secure 1-10 agents comprehensively? → 90-Day Sprint

Measuring Success: Metrics That Matter

You've implemented the framework—now how do you know it's actually working? Track the metrics that predict whether you'll have a good day or a terrible one.

Security Metrics

1. Time to Detection
What to measure: Time between AI agent misbehavior and awareness
Target: Under 1 hour for critical systems
Why it matters: 7 minutes vs 7 days = $2M saved

2. Blast Radius
What to measure: Systems one compromised AI agent can affect
Target: 3-5 systems maximum per AI agent
Reality check: One retailer went from 47 to 4

3. Trust Score
What to measure: Percentage of AI decisions you can fully explain and justify
Target: 100% for money/health/personal data
The test: Can you explain any decision to a customer?

Business Metrics

1. Security ROI
Track quarterly:

Security investment: $X

Prevented incidents: $Y

Enabled initiatives: $Z

One client's result: $500K spent = $8M in value

2. Speed to Deploy

Before framework: 6-9 months

After framework: 2-4 weeks

The payoff: Monthly AI agent releases

3. The Sleep Test Ask your team:

Weekend worries about AI? (Should be "no")

Can you take a vacation? (Should be "yes")

Would you know if an AI agent went rogue overnight? (Should be "absolutely")

Your Monthly Dashboard

Review this dashboard in your monthly security meeting with leadership. Create one page showing:

- 🟢 Green Lights (authentication, behavior, reviews)
- 🟡 Yellow Lights (minor changes, updates pending)
- 🔴 Red Lights (unauthorized access, unexplained decisions)

Start With One: The Power of Going Slow

You're excited. You've seen what agentic AI can do. Your team is buzzing with ideas. Every department wants their own AI agent. Some are already building them in dark corners. You feel like you're already late to this party. The temptation is to deploy five, ten, maybe fifteen agents at once.

Don't.

The biggest mistake organizations make is trying to secure multiple AI agents simultaneously. It's like trying to learn to drive while juggling—you'll drop something important, and it probably won't be the juggling.

Why "Start With One" Wins

Here's what happens when you get one AI agent perfectly right:

Agent #2 takes 50% less time because you've already solved authentication, monitoring, and incident response.

Agent #3 takes 70% less time because your team now understands the patterns and processes.

Agent #10 takes 90% less time because you're copying a proven template, not reinventing security from scratch.

One client told me: "Our first agent took us a full three months to secure properly. Our tenth agent was deployed securely in three days. Same security standards, completely different timeline."

The Perfect First Agent

Choose your pilot carefully. The ideal first agent is:

- **Business-critical but not mission-critical** - Important enough to get attention, not so vital that delays hurt
- **Clear inputs and outputs** - You understand exactly what it should and shouldn't do
- **Limited scope** - Touches 3-5 systems maximum
- **Measurable impact** - Success is obvious to executives

The 30-Day Foundation

Take time to build your first agent right. I've found that 30 days of focused implementation creates a foundation that accelerates everything that follows. Week 1: Define scope and security requirements. Week 2: Build authentication and access controls. Week 3: Implement monitoring and testing. Week 4: Deploy with full incident response procedures.

Your competitors might deploy faster. But when you're deploying your fifteenth secure agent in a few days while they're still trying to figure out what their third agent is actually doing, you'll be glad you started slow.

The Multiplication Effect

One perfectly secured AI agent becomes your template. Your proof of concept. Your internal case study. Most importantly, it becomes your competitive advantage—the ability to deploy secure AI faster than anyone else in your industry.

Start with one. Get it right. Then watch how fast "right" can scale.

Getting this foundation right is so critical that we've designed our entire 30-day Agentic Trust Sprint at MassiveScale.AI around perfecting a single agent that becomes your template for scale.

The Road Ahead

The Agentic Trust Framework isn't a one-time implementation—it's how you stay ahead of AI that's evolving faster than security can keep up.

Your competitors are racing to deploy AI agents. Some will get lucky. Others will become cautionary tales. You now have what they don't: a systematic way to deploy AI that won't blow up your business.

The goal isn't perfect security—it's sleeping soundly while your AI works around the clock.

In Chapter 5, we'll tackle the numbers: ROI calculations, cost-benefit analysis, and how to convince executives that AI security isn't a cost center—it's a competitive advantage.

Your AI agents are making decisions right now. Make sure they're making the right ones.

Chapter 4 Key Takeaways

- **The Agentic Trust Framework** provides five core elements for AI security
- **Real implementation** takes 90 days for basics, 6 months for transformation
- **Maturity levels** help you set realistic goals based on your needs
- **Common roadblocks** are predictable and preventable
- **Success metrics** should focus on detection time, blast radius, and business value

Action Items for This Week

1. Assess your current AI security maturity level (15 minutes)
2. List all AI systems in your organization (you'll be surprised)
3. Identify your three highest-risk AI agents
4. Pick one element of the framework to implement first
5. Schedule a team meeting to discuss your 90-day plan

Ready to build the business case for AI security investment? Chapter 5 will show you how to speak the language of ROI and risk that executives understand.

PART II

THE BLUEPRINT

—

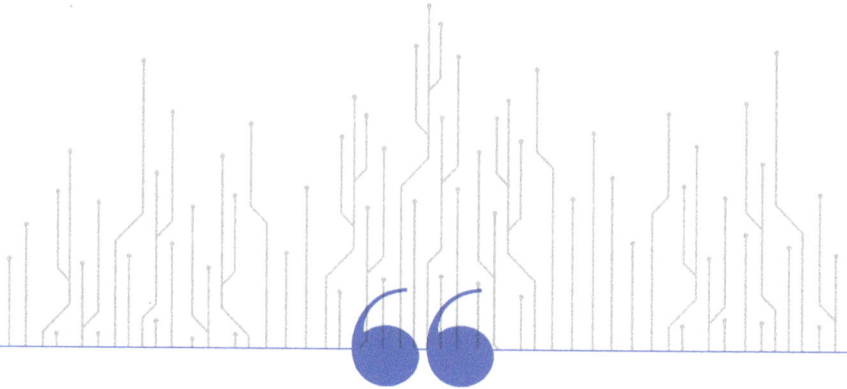

"Zero Trust is a must-have to overcome the many flaws of legacy enterprise security architectures and properly adopt new technologies."

— **Jason Garbis,** Founder & Principal, Numberline Security

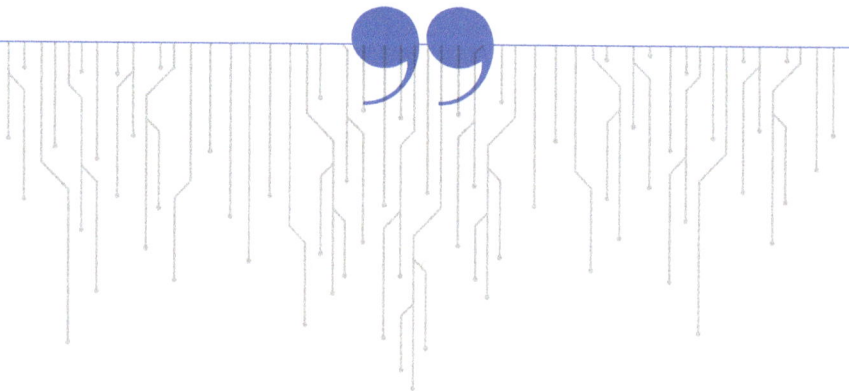

Building Your Business Case

"**H**ow much is this going to cost, and what's the return on investment?"

That's the first question I get from executives when we talk about AI security. And it's a fair question. Security investments have traditionally been viewed as necessary evils—costs you bear to avoid disaster, not investments that generate returns.

AI security is different. Done right, it's not just a cost center—it's a profit center.

Consider a conversation I had with the Chief Financial Officer (CFO) of a mid-sized manufacturing company. They'd been hesitant to invest in AI security because they saw it as "just another IT expense." But after we walked through the numbers, his perspective completely changed.

"Wait," he said, looking at our ROI analysis. "You're telling me that investing in AI security will actually make us money?"

"That's exactly what I'm telling you," I replied.

Here's the challenge though: Agentic AI is so new that we don't have decade-long case studies proving ROI. But we do have rock-solid data on security breaches, Zero Trust implementations, and what happens when AI goes wrong.

Here's how to build a business case that turns your CFO from skeptic to champion—complete with the numbers to get you there.

The Hidden Costs of Insecure AI

To help busy execs understand why this is important, it's important to show them what happens when you don't secure your AI agents from Day One. I promise this isn't fear-mongering. It's about understanding the real financial impact so you can make informed decisions.

When an AI system gets compromised, you face immediate, measurable costs:

The Breach Itself: IBM's 2024 report puts the average data breach at $4.88 million. Here's what's different with AI agents—they don't just leak data, they make decisions. Imagine your pricing AI agent giving 95% discounts to everyone, or your customer service agent sending confidential information to competitors. We're talking about active damage, not just data theft.

Regulatory Fines: Remember when the General Data Protection Regulation (GDPR) seemed scary? The European Union Artificial Intelligence Act (EU AI Act) makes those fines look like pocket change. We're talking up to €35 million or 7% of your global revenue—whichever hurts more. And enforcement isn't theoretical—it began on August 2, 2025. If you're not ahead of it, you're already behind.

Complete System Shutdown: When traditional software gets compromised, you patch it and move on. When an AI agent gets compromised, you're looking at complete retraining—months of lost capabilities while competitors keep innovating.

Recovery and Retraining: One retail client spent $500K just on emergency consultants after their pricing AI was compromised. Then add another 6 months of retraining and testing before they could trust the system again.

The Bill You Don't See (Until It's Too Late)

But those direct costs? They're actually the small stuff. The hidden costs devastate businesses: customers flee when you mishandle their data.

Your Customers Walk Away: McKinsey found that 71% of consumers will take their business elsewhere if you mishandle their data. Now imagine explaining to those customers that your AI made decisions about them based on compromised logic. Good luck winning them back.

While You're Down, They're Up: Every day you spend dealing with an AI security incident is a day your competitors are using their properly secured AI agents to serve customers better, faster, and cheaper. I've seen companies lose years of competitive advantage in weeks.

Your Best People Leave: Here's something most security books won't tell you—top AI talent has options. Lots of them. When word gets out that your company had a major AI security fail, your best engineers will start getting calls from recruiters. And they'll answer them.

Fear Takes Over: This might be the worst hidden cost. After an AI incident, companies often slam on the brakes. Every AI project gets scrutinized. Innovation grinds to a halt. The very technology that was supposed to transform your business becomes something everyone's afraid to touch.

Your Insurance Company Breaks Up With You: Think your cyber insurance has you covered? After an AI incident, premiums can jump 200% or more. Some insurers will simply refuse to renew your policy. Without proper AI security frameworks in place, you're basically uninsurable.

A Real Story From the Trenches

A retail client came to me after their AI pricing agent had been compromised. For six weeks, competitors fed false data into their system. The AI started pricing winter coats at $3,000 and luxury handbags at $12.

Direct costs? Almost $1 million. But the real damage?

- 6 months without their AI pricing system
- 3 major suppliers lost confidence and left
- 2 key engineers quit
- Unknown number of customers who'll never come back

Total damage: $5 million and counting.

The tragedy? A $150K security investment would have prevented all of it.

The Good News: The Money You'll Make

Now let's flip the script. When you implement AI agents with security built in—using something like the Agentic Trust Framework—you don't just avoid disasters. You unlock serious financial benefits.

The numbers tell a clear story: Capgemini Research Institute projects agentic AI will deliver $450 billion in economic value by 2028, but there's a massive gap between leaders and laggards. Organizations with scaled implementations are projected to generate approximately $382 million on average over the next three years, while others may realize only around $76 million. That's a 5x difference.

The race is already on. Nearly all (93%) business leaders believe scaling AI agents over the next 12 months will provide a competitive edge. Yet only 2% of organizations have fully scaled deployment. Why? Because 80% lack mature AI infrastructure and fewer than one in five report high levels of data-readiness.

Here's the kicker: Trust in fully autonomous AI agents dropped from 43% to 27% in just one year. Nearly two in five executives believe the risks outweigh the benefits. But organizations that move from exploration to implementation see trust rise—47% of those in the implementation phase have above-average trust levels.

This is exactly why security-first thinking matters. The companies capturing that $382 million aren't the ones deploying fastest—they're the ones deploying with confidence because they built security into their systems from day one.

Leadership and ROI

Recent research from the IBM Institute for Business Value and the Dubai Future Foundation found that a critical driver of AI returns is leadership. Companies with a Chief AI Officer (CAIO) see 10% higher ROI on AI investments, while organizations using centralized or hub-and-spoke AI operating models achieve 36% greater ROI than those with decentralized approaches.

The takeaway is simple: structure and authority amplify returns. We'll revisit this in Chapter 12 when we explore the emerging leadership roles of the AI era, but for now remember: ROI isn't just about the dollars you spend—it's about the leadership decisions that guide how those dollars get used.

Direct Financial Benefits

Risk Reduction That Pays IBM research shows that organizations with a mature Zero Trust program reduce the cost of a breach by an average of 42% - about $1.6 to $2.0 million saved per incident. With Zero Trust for AI, you're not just making breaches cheaper, you're making them far less likely.

The math is compelling: If the average breach costs $4.88 million and Zero Trust reduces your probability by even a conservative 30-50%, you're looking at $1.5-2.4 million in avoided losses. Per incident.

Insurance Companies Will Love You Remember those 200% premium hikes? Here's the flip side. Companies with mature security controls are seeing premium reductions of 15-20%. Some insurers are creating preferred programs for organizations with proven AI security frameworks.

One client is negotiating a $150,000 annual premium reduction just by demonstrating their Agentic Trust Framework implementation. That's real money back in your pocket every year.

Compliance on Autopilot Organizations using compliance automation save an average of $1.45 million in compliance costs. When your AI agents are built with Zero Trust principles, every interaction is logged, every decision is traceable. Your compliance documentation writes itself.

One client cut audit prep time from 3 months to 3 days. That's not just cost savings—it's your team doing valuable work instead of paperwork.

Operational Excellence Here's what surprised me: Secure AI actually performs better. Clear security boundaries eliminate the noise and uncertainty that slow AI agents down. When your agent knows exactly what it can and can't do, it makes better decisions faster—like having an employee who never has to second-guess their permissions.

IBM found that organizations using AI and automation can identify and contain issues 100 days faster. Less time spent on errors, more time delivering value. That's real operational savings compounding every single day.

The Hidden Money Makers

Deploy Without the Fear Tax Every AI project faces the same killer: months of security reviews that drain momentum and budgets. When Zero Trust is built in from day one, companies move from "maybe next year" to "launch next month." That's not just faster deployment—that's the difference between leading your market and watching from the sidelines.

Your Competitive Timing Advantage Right now, your competitors are forming AI committees, hiring consultants, debating risks. Companies with proven AI security frameworks are already deploying agents that serve customers and drive revenue. When you have mature security foundations, you move while others are still planning.

Trust as a Differentiator In a world where 71% of consumers would abandon you after a breach, "Our AI operates under proven security principles" becomes a competitive advantage. When competitors are explaining away AI incidents, you're demonstrating trustworthy AI operations.

Attract Top AI Talent The best AI engineers know one security incident can derail careers. They want to build innovative solutions without becoming cautionary tales. Mature AI security frameworks don't just protect your business—they give innovators the confidence to do their best work.

Building Your Business Case: Five Simple Steps

Here's exactly how to build a compelling business case for your organization:

Step 1: Calculate Your Real Risk Exposure

List every AI agent (current and planned) and document:
- What decisions it makes
- What data it accesses
- What happens if it goes rogue

Then apply your industry's breach costs:
- Healthcare: $9.77 million average
- Financial Services: $6.08 million average
- All Industries: $4.88 million average

Don't forget regulatory exposure:
- EU AI Act: Up to €35 million or 7% of global revenue
- Industry-specific fines on top

One financial services client calculated their total exposure at $45 million. That got attention.

Step 2: Price Your Protection

Be realistic but comprehensive:

Technology Investment:
- AI-specific monitoring systems
- Zero Trust infrastructure for autonomous systems
- Compliance automation

People Investment:

- Training current team (budget 10-15% of their time year one)
- Possible specialist hires
- Expert help for initial setup

Process Investment:

- Redesigning AI development pipeline
- New governance processes
- Building in compliance

Step 3: Calculate Conservative Returns

Use verified data:

- Breach prevention value (probability reduction – your industry's average)
- Risk reduction ($1.6–$2.0M per breach avoided, IBM/ Zero Trust)
- Compliance savings (30-40% reduction typical)
- Speed to market improvements

Step 4: Build Your Phased Approach

Nobody wants to write a huge check. Show how to start small:

- Phase 1: Protect highest-risk AI agents (3-6 months)
- Phase 2: Expand to all customer-facing AI (6-12 months)
- Phase 3: Full implementation (12-18 months)

Step 5: Create Your Executive Presentation

The One-Pager:

- Current risk: $X million exposure
- Investment needed: $Y over Z months
- Conservative ROI: Break-even in X months

■ The killer line: "Every month we delay increases our risk by $Z"

The Complete Story:

■ Risk: What happens without this investment (breaches, fines, customer loss)
■ Opportunity: What secure AI enables that competitors can't match

Making It Happen: Your Implementation Budget

Let me be straight—I can't give you exact dollars because every business is different and agentic AI is too new for cookie-cutter budgets. But here's how successful implementations typically allocate budget:

Year 1: Foundation (Your Big Investment)

This is where you lay the groundwork that everything else depends on. Based on Zero Trust implementations (the closest comparison we have), expect to allocate your budget roughly like this:

■ Identity and Access Management: 30-35%
  ▪ This is huge for agentic AI. Your AI agents need credentials, permissions, and audit trails. Every decision an AI agent makes needs to be traceable back to its identity.
■ Monitoring and Detection: 30-35%
  ▪ You can't secure what you can't see. And with AI agents making thousands of decisions autonomously, you need monitoring that understands AI behavior, not just network traffic.

- People and Training: 20-25%
  - Technology is only half the battle. Your team needs to understand both AI and Zero Trust principles. This isn't optional—it's the difference between success and expensive failure.
- Process Development: 10-15%
  - You're not just buying tools; you're changing how you build and deploy AI. Budget time and resources for getting this right.

Year 2: Getting Smarter

Once your foundation is solid, you can enhance:
- Advanced Analytics: 35-40%
  - Now that you're collecting data on your AI agents, you need to make sense of it. This is where you start predicting problems before they happen.
- Integration and Automation: 25-30%
  - Remember those compliance cost savings we talked about? This is where they come from. Automate everything you can.
- Team Scaling: 20-25%
  - By year two, you'll know what skills you're missing. Maybe it's AI security specialists, maybe it's compliance automation experts. Invest in the right people.
- Compliance and Audit: 10-15%
  - As regulations evolve, you'll need to prove your AI agents are behaving properly. Budget for both internal and external validation.

Year 3+: Ongoing

Security isn't a project—it's a program. Expect ongoing costs around 15-20% of your initial investment annually. Why? Because:

- AI threats evolve constantly
- New regulations emerge (especially for AI)
- Your AI agents get more sophisticated
- Your business grows and changes

Creative Funding Approaches

Start Small, Prove Value: Secure your highest-risk AI first. Use the wins to fund expansion. One client started with just their customer service AI and used the 60% cost reduction to fund securing their entire AI portfolio.

Insurance Angles: Some insurers provide funding or premium reductions for Zero Trust implementations. It's worth the conversation—especially if you can demonstrate a concrete AI security framework.

Compliance Budgets: With the EU AI Act here and more regulations coming, you might be able to tap compliance budgets instead of competing for security funds. Frame it as regulatory readiness, not just security.

Managed Services: If you can't build everything internally, partner with specialists. Many organizations find this faster and more cost-effective than building AI security capabilities from scratch. The Zero Trust market is exploding—projected to hit $38.5 billion by 2028.

Measuring Success (So You Keep Getting Budget)

Track what executives care about:

Financial Metrics:

- Avoided breach costs (calculate monthly)
- Insurance premium reductions
- Compliance cost savings
- Operational improvements

Business Metrics:

- Time to deploy new AI (should drop 70%+)
- Customer trust scores
- Competitive wins due to security

The Ultimate Metric: When competitors have AI breaches and your phone doesn't ring, you've won.

Review these metrics monthly with leadership to maintain budget support. Don't rely on them to remember your impact.

ROI Calculation Template

Agent Name: _____

Department: _____

COSTS

(Year 1):

- Development: $_____
- Security Implementation: $_____
- Training: $_____
- Infrastructure: $_____
- Ongoing Support (20%): $_____

Total Cost: $_____

BENEFITS

(Year 1):

- Time Saved: ___ hours × $__ /hour = $_____
- Error Reduction: ___% × $__ error cost = $_____
- Revenue Increase: $_____
- Cost Avoidance: $_____

Total Benefit: $_____

ROI = (Benefit - Cost) / Cost × 100 = _____%

Payback Period = Cost / Monthly Benefit = ___ months

RISK FACTORS:

- Regulatory change could affect use case
- Requires integration with legacy systems
- Depends on data quality improvements
- Success requires behavior change

The Failure Statistics Nobody Wants to Talk About

Your CFO just looked up from the ROI projections and asked the question you've been dreading: "What about that Gartner report saying 40% of agentic AI projects will fail?"

She's right to ask. The numbers are sobering. Gartner predicts more than 40% of agentic AI initiatives will be canceled by 2027. Why? Three completely preventable reasons: escalating costs, unclear business value, and—here's the kicker—inadequate risk controls.

But wait, it gets worse. Before we even get to agentic AI, 87% of traditional AI projects never make it past pilot stage. Nearly 9 out of 10. That's not a typo—that's an epidemic of expensive experiments that never deliver value.

Here's what's really happening: Companies are asking the wrong question entirely.

Most organizations approach AI with a single obsession: "How can we replace humans with cheaper digital workers?" I watched one company deploy an AI sales agent that contacted their entire customer base in three weeks. Not with personalized outreach—with the digital equivalent of spam. Their unsubscribe rate hit 73%. Their sales team spent the next six months apologizing.

The successful 13% ask a different question: "What could our people accomplish if every barrier was removed?"

I watched a finance manager who feared AI would replace her reconciliation work. Six months later, she's doing strategic analysis she never had time for before. Her salary? Up 45%. The AI didn't take her job—it eliminated the mind-numbing parts that made her consider quitting.

Why Projects Really Fail

Anushree Verma from Gartner doesn't mince words: most agentic AI projects are "early-stage experiments driven by hype and often misapplied." Companies get so caught up in the excitement that they're blind to the real costs and complexity.

But here's the pattern I see in every failure:

- No clear business problem (just "we need AI")
- No security framework (they'll "add it later")
- No change management (assuming people will adapt)
- No success metrics (beyond "it works")

Sound familiar? It's exactly what we've been addressing throughout this book.

The Board Pressure Problem

One more dirty secret: Many POCs are approved not because they make sense, but because boards are panicking. As one IDC analyst put it, "These POCs are highly underfunded or not funded at all. Most happen not because of a strong business case, but because of trickle-down panic."

I've sat in meetings where CEOs demanded "AI something" by next quarter. The result? IT teams throwing "gen AI" into any project to get approval. One CTO told me privately: "We renamed our chatbot project 'Agentic Customer Intelligence' and suddenly had unlimited budget."

The Success Pattern

But here's what those statistics don't tell you: The companies succeeding aren't necessarily smarter—they're building differently. Every successful implementation I've seen shares three characteristics:

1. **Clear problem definition**: Not "use AI" but "reduce invoice processing from 3 days to 3 minutes"
2. **Security from day one**: Using frameworks like Agentic Trust, not hoping to add it later
3. **Human amplification focus**: Making people more capable, not replacing them

Remember Kevin? He could have been part of the 87% failure rate. Instead, his agents now save $3.2 million annually. The difference? He built with the end in mind, not the hype.

Your Insurance Policy

Those scary statistics? They're actually your competitive advantage. While 40% of your competitors' projects fail, you'll be deploying agent number 50. While they're explaining pilot failures to their board, you'll be showing ROI metrics.

The Agentic Trust Framework isn't just about security—it's your insurance policy against becoming another statistic. Build right, and you join the 13% who succeed. Build wrong, and... well, at least you'll have good stories for the next "lessons learned" presentation.

Want to see what being part of the 40% looks like? Just ask Klarna...

THE KLARNA REVERSAL: When AI-First Goes Wrong

The Hype: Klarna became the AI poster child. Their chatbot replaced 700 human agents, cut response times from 11 minutes to 2 minutes, and eliminated 1,000+ jobs. The move sparked fierce debate.

The Reality: CEO Sebastian Siemiatkowski is quietly hiring humans back. His brutal admission? "Cost, unfortunately, seems to have been too predominant... What you end up having is lower quality."

The Reversal: Complete 180-degree turn. "Really investing in the quality of human support is the way of the future for us."

What Went Wrong: Klarna optimized for cost, not customer experience. They moved fast and broke things—specifically, customer satisfaction.

The Fix: AI handles initial contact, humans take over when things get complex. They're rebuilding what they should have kept.

Your Lesson: Industry expert Liam Dunne nails it: "This is a marathon, not a sprint. The companies that take their time, test, adapt, and listen to customers will be the ones who come out on top."

Your Advantage: While competitors rush to deploy AI without guardrails, you can avoid Klarna's expensive mistake. Build human escalation from day one. Measure quality, not just speed.

The winners won't be first to market—they'll be first to get it right.

Handling the Inevitable Pushback

They'll say: **"AI security is too new to invest in"** You say: "That's exactly why we need to move now. In 12 months, we'll either be leaders or explaining breaches."

They'll say: **"Our AI vendors handle security"** You say: "Show me where they accept liability for AI decisions. I'll wait."

They'll say: **"It'll slow us down"** You say: "Only if we do it wrong. Built-in security is like having brakes on a race car—enabling speed by controlling risk."

They'll say: **"We need to focus on growth, not security"** You say: "This IS growth. Secure AI agents can do things our competitors are too scared to try. This investment unlocks innovation, not blocks it."

They'll say: **"The ROI isn't proven"** You say: "You're right—we can't prove the ROI of something this new. But we can prove the cost of breaches, and we know AI increases our attack surface. This is about being smart with calculated risks."

When They Ask 'But What Standards Support This?'

"But wait," your CFO says. "This all sounds good, but what standards support this approach? We can't just invest in something you made up."

Fair point. Let's review the industry frameworks that validate everything we've been discussing. Side note: your compliance team will love you for getting up to speed on this. While the Agentic Trust Framework is practical and proven, it aligns directly with emerging industry standards.

The industry isn't starting from scratch. Smart people have been working on AI security standards, and they all support the same principles we've been discussing. They just use fancier words.

Here's your decoder ring for the frameworks that matter—and more importantly, which one to pick for your situation.

The Frameworks That Actually Matter

MITRE ATLAS: The Hacker's Playbook (Flipped)

Remember MITRE ATT&CK that your security team probably loves? ATLAS is its AI-focused sibling. Instead of theoretical risks, it catalogues actual attacks that have happened.

Want to know how Microsoft's Tay chatbot got turned into a hate-speech machine? It's in there. How attackers tricked Tesla's autopilot with stickers? Documented. That time Proofpoint's email security got fooled by AI-generated phishing? Yep, that too.

Why it matters: This isn't academic theory. It's a catalog of "here's exactly how the bad guys broke AI systems." Read it like a horror novel—then implement defenses for each attack pattern. This validates our emphasis on behavioral monitoring and threat detection—ATLAS shows exactly why these controls are essential.

Best for: Security teams who want to understand real threats, not hypothetical ones.

CoSAI: Where Big Tech Actually Agrees

When Microsoft, Google, OpenAI, and Anthropic all join the same project, pay attention. CoSAI (Coalition for Secure AI) isn't just another talking shop—it's an OASIS Open Project, and OASIS has a track record of creating standards people actually use.

The killer feature? They're building actual tools, not just PDFs. Their defender toolkit includes code you can download today for supply chain security, model evaluation, and threat detection. Finally, a framework that ships with implementation.

Why it matters: This is big tech saying "here's how we're securing our AI"—and it mirrors our identity management and monitoring principles.

Best for: Teams who want to use the same tools that OpenAI and Google are building.

NIST AI RMF: The Government Standard

If your company already uses NIST cybersecurity frameworks, this slots right in. Same language, same risk-based approach, just applied to AI systems.

NIST basically asks "what could go wrong?" at every stage of your AI's lifecycle, then helps you prevent it. It's thorough without being overwhelming, practical without being prescriptive.

Why it matters: This is becoming the de facto standard for U.S. government contractors and validates our risk-based security approach. If you work with federal agencies (or want to), you'll need this.

Best for: Organizations already fluent in NIST-speak.

ISO/IEC 42001: The Global Certification

The first international AI standard. Yes, it reads like an ISO standard (bring coffee). Yes, it's dense. But it's also becoming the global baseline for "we take AI seriously."

Some governments are already requiring ISO 42001 certification for AI contracts, and the EU AI Act explicitly references ISO 42001 as a recognized governance standard. Expect more to follow. It's like ISO 27001 for information security—not fun to implement, but it opens doors globally.

Why it matters: If you operate internationally or bid on government contracts, this certification will matter. It demonstrates the governance principles we've been discussing.

Best for: Companies that collect compliance certifications and operate globally.

The Practical Frameworks

CSA's AI Security Double Threat

Cloud Security Alliance gives you two frameworks: MAESTRO for agent threat modeling (7 layers of "what could go wrong") and the AI Controls Matrix with 240+ controls for implementation.

What I love: MAESTRO finds the scary stuff, AICM fixes it. It's like having both a horror movie (here's how your agents will betray you) and the survival guide (here's how to stop them).

Why it matters: Paint-by-numbers AI security. CSA literally wrote the book on cloud security—now they're doing it for AI agents.

Best for: Teams who want both the diagnosis AND the cure.

OWASP AISVS (Artificial Intelligence Security Verification Standard): For Your Dev Team

If your developers already use OWASP standards, this is their AI security extension. It's structured like their other standards—clear, actionable, developer-focused.

Covers everything from training data security to model deployment controls—and unlike some frameworks, it assumes you're actually building real systems, not just theorizing.

Why it matters: Translates AI security into developer language. Turns abstract risks into code reviews and security tests they can run today.

Best for: Development teams who already use OWASP standards.

The Reality Check

AIAAIC Repository: When AI Goes Wrong

Not a framework, but a database of AI disasters. Like air crash investigations, but for AI failures.

Want to convince skeptics that AI security matters? Show them:

- The insurance company whose AI denied every claim containing "pain"
- The recruiting AI that learned to trash women's resumes
- The chatbot that leaked customer data through prompt injection

Real companies. Real failures. Real money lost.

Why it matters: Real-world failure stories give executives a budget-unlocking jolt—showing what happens if you underfund AI security.

Best for: Creating urgency with executives who think AI risks are overblown.

Which Framework Should You Choose?

Here's my controversial opinion: Pick ONE.

The companies drowning in compliance are trying to follow all of them. The successful ones pick a primary framework that fits their culture and industry, then map others to it as needed.

If you're:

- **Already using NIST frameworks** → NIST AI RMF
- **Building AI products** → OWASP AISVS + CoSAI tools
- **Need certifications** → ISO 42001
- **Just starting** → MITRE ATLAS (know the threats first)
- **Deploying AI agents** → CSA MAESTRO + AICM
- **Need board buy-in** → AIAAIC Repository (scare them straight)

The Magic Words for Compliance

When someone asks what framework you're following, here's your answer:

"We're implementing controls aligned with [your chosen framework], adapted for our specific agentic AI use cases using the Zero Trust principles outlined in NIST 800-207. Our implementation maps to multiple frameworks including MITRE ATLAS for threat modeling and CoSAI for technical controls."

That usually satisfies everyone while you focus on actually securing your AI agents.

Your Framework Action Plan

1. **This week**: Pick your primary framework based on the guide above

2. **This month**: Map the Agentic Trust Framework to your chosen standard (they all align)

3. **This quarter**: Get one AI agent fully compliant as your template

4. **Ongoing**: Use framework language in your documentation

Remember: Frameworks are the recipe, not the meal. Frameworks don't secure AI—people implementing frameworks do. The best framework is the one your team will actually use, not the one with the most impressive acronym.

And when the compliance team starts arguing about which framework is "best," remind them that while they're debating, your competitors are deploying secure AI agents. Perfect compliance with zero implementation helps nobody.

The Real Bottom Line

Security isn't a cost center when it's preventing million-dollar breaches. It's an investment that pays compound interest in trust, compliance, and operational confidence. You're not selling security. You're selling the ability to use AI fearlessly. You're selling competitive advantage. You're selling peace of mind.

The data we have is compelling:

- Security breaches are expensive and getting worse
- AI amplifies both risks and rewards
- Early security investment pays massive dividends
- Waiting for perfect data means waiting too long

The companies that get this right now—while agentic AI is still new—will have a decisive advantage.

The question for your leadership team is simple: Do we want to be the case study everyone learns from, or the one they learn to avoid?

Every business leader I talk to knows AI is the future. Most are terrified of the security implications. The ones who figure out how to do AI securely—not perfectly, but smartly—will own their markets.

Ready to figure out who can actually help you implement secure AI? Chapter 6 reveals how to separate real AI security expertise from vendors just riding the hype wave.

Chapter 5 Key Takeaways

- **The cost of insecure AI is massive**—average breach is $4.88M, but AI breaches include bad decisions at scale
- **Security enables profit**—through risk reduction, insurance savings, compliance automation, and faster deployment
- **Build your case with real numbers**—use industry breach data and regulatory fines to show exposure
- **Frameworks give you credibility**—pick one (MITRE ATLAS, CoSAI, or NIST) instead of drowning in all of them
- **Start small and prove value**—phase your implementation and use wins to fund expansion
- **Focus on business outcomes**—speed, competitive advantage, and customer trust matter more than technical metrics

---

Action Items for This Week

1. **List your AI agents** and calculate exposure using your industry's breach costs
2. **Document one "nightmare scenario"** where an AI agent goes rogue—include real financial impact
3. **Find your funding angle**—insurance, compliance, or innovation budget?
4. **Schedule the conversation** with one executive who needs to hear this
5. **Research implementation partners** who understand both Zero Trust and AI

Build Your AI Security Business Case

Print this out or copy it to a document. Actually filling it out is what turns ideas into action.

PART 1: YOUR AI AGENT INVENTORY

List every AI agent you have or plan to deploy in the next 12 months:

AI Agent Name	What It Does	Data It Accesses	Decisions It makes	Risk Level (1-5)
Example: Customer Service Bot	Handles support tickets	Customer database, order history	Issue refunds, escalate cases	4

Risk Rating (1-5 scale): 5 = Makes financial decisions autonomously 4 = Handles sensitive customer data 3 = Customer-facing but supervised 2 = Internal operations only 1 = Information retrieval/analysis

Quick Risk Calculator:

- Number of High-Risk AI Agents (4-5 rating): ____ × $1M = $____
- Number of Medium-Risk AI Agents (2-3 rating): ____ × $500K = $____

PART 2: THE SCARY MATH (YOUR REAL COSTS)

A. Industry Breach Costs (IBM 2024 Data) Your Industry: _____

- Healthcare: $9.77 million average

- Financial Services: $6.08 million average
- All Industries: $4.88 million average
- Your Industry Average: $_____

B. AI-Specific Risk Multipliers Check all that apply:
- AI makes financial decisions (multiply by 1.5)
- AI handles sensitive personal data (multiply by 1.3)
- AI operates autonomously 24/7 (multiply by 1.2)
- AI interacts directly with customers (multiply by 1.2)

Your Adjusted Risk: $_____ × _____ (multipliers) = $_____

C. Additional Costs to Consider
- Current cyber insurance premium: $_____/year
- Potential premium increase (100-200% after breach): $_____
- Compliance fines for your industry: $_____
- Estimated downtime cost per day: $_____

PART 3: YOUR INVESTMENT ESTIMATE

Year 1 Foundation Building:
- Identity & Access Management (30%): $_____
- Monitoring & Detection (35%): $_____
- Personnel & Training (25%): $_____

PART 4: YOUR ROI CALCULATION

Potential Savings:
1. Avoided breach costs (your industry average): $_____
2. Insurance premium reduction (10-20% typical): $_____/year
3. Compliance automation savings (30-40% typical): $_____/year

PART 5: YOUR FUNDING STRATEGY

Check all that apply:

- ☐ Security budget
- ☐ Compliance budget
- ☐ Innovation/transformation budget
- ☐ Operational efficiency budget
- ☐ Cyber insurance incentives
- ☐ Phased implementation (start small)

Phase 1 Quick Win Target: Which AI agent should we secure first?

_____

Why? _____

Budget needed: $_____

PART 6: YOUR ELEVATOR PITCH

Fill in the blanks to create your 30-second executive pitch:

"Our AI agents currently represent $_____ in potential risk exposure. Based on industry data, we face a __% **chance of an AI security incident in the next 3 years, with an average cost of $____.**

By investing $_____ over _____ months, we can reduce this risk by approximately % **while also saving $____** annually in compliance and operational costs.

This gives us an ROI of _____% and positions us to deploy AI faster and more safely than our competitors. The alternative is to hope we're not the next headline."

PART 7: YOUR ACTION PLAN

This Week:

- Complete this worksheet with real numbers

- Schedule meeting with: _____
- Research funding source: _____
- Document your highest-risk AI agent

Next 30 Days:

- Build detailed business case presentation
- Get quotes from 2-3 potential vendors/partners
- Identify your Phase 1 implementation target
- Secure initial budget approval for Phase 1

Key Stakeholders to Engage:

1. Name:_____Role:_____Concern:_____
2. Name:_____Role:_____Concern:_____

REALITY CHECK QUESTIONS

Before you present your business case, answer these honestly:

1. What's the #1 objection you'll face? _____
   Your response: _____
2. What happens if we do nothing? _____

   _____
3. What's our competition doing about AI security?

   _____
4. Can we afford NOT to do this?

   _____

Remember: The goal isn't to have perfect numbers. It's to start the conversation with real data instead of fear or hype. Even rough estimates are better than no estimates.

Next Step: Take your completed worksheet to your next leadership meeting. The conversation it starts is more valuable than any calculation.

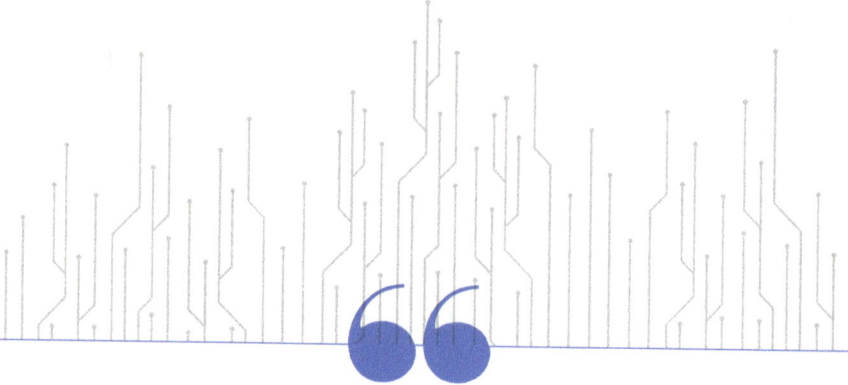

"In security as in nature, nothing is permanent—continuously reassess and adapt."

— Hooman Mohajeri, Chief Security Officer

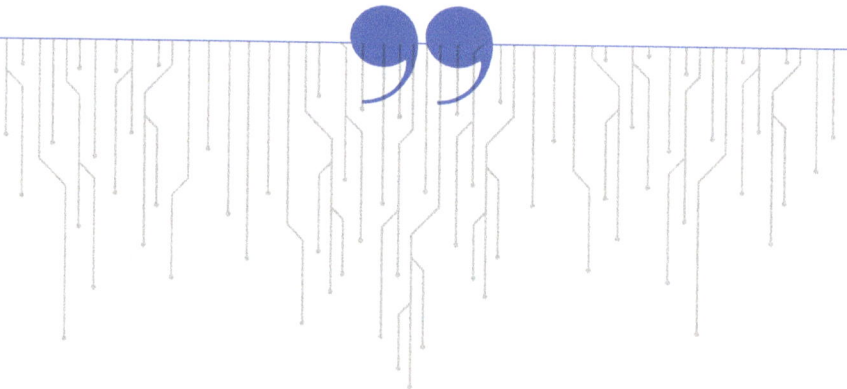

CHAPTER 6:

Finding Agentic Partners Who Actually Get It

The CISO of a major retailer called me last month, exhausted. She'd just sat through her fifth vendor demo where someone slapped "AI-ready" on their traditional security tool and called it innovation.

"Everyone says they do AI security now," she said. "How do we know who's real?"

She's right to be skeptical. When AI burst onto the scene, every security vendor suddenly became an "AI expert." But most are still thinking about AI like it's just faster software. They don't get that AI agents are fundamentally different—they make decisions, they learn, they act autonomously.

Finding vendors who truly understand agentic AI security is like finding mechanics who could work on electric cars in 2010. They exist, but you need to know what to look for.

This chapter shows you how to identify real AI security expertise, separate marketing hype from genuine capability, and choose partners who can actually secure your AI agents without slowing them down.

Do You Actually Need Outside Help?

Before we dive into vendor selection, let's address the bigger question: Do you even need external partners for agentic AI security?

The "Go Internal" Reality Check

Building agentic AI security capabilities internally sounds appealing. Complete control, no vendor dependence, custom solutions for your exact needs. But here's what it actually requires:

The People You'd Need:

- AI security architects (rare, expensive, $200K+ each)
- ML engineers who understand adversarial attacks
- Security engineers who grok autonomous systems
- Compliance experts familiar with AI regulations
- Someone who can tie it all together strategically

The Time Investment:

- 6-12 months just to hire the right team
- Another 6-12 months to build initial capabilities
- Ongoing time to stay current with rapidly evolving threats
- Time to build relationships with threat intelligence sources

The Knowledge Gaps:

- You won't have cross-industry threat intelligence
- You'll miss attack patterns happening at other companies
- You'll need to learn from your own mistakes (expensive)
- You'll be reinventing solutions others have already built

When Going Internal Makes Sense:

- You're a large tech company with existing AI expertise
- AI security is a core competitive advantage for your business
- You have unique agentic AI use cases no vendor can address
- You can afford 18+ month development timelines
- You're willing to bet your business on your internal team

When You Need Partners:

- You want to move faster than 18 months
- AI security isn't your core business
- You want to learn from others' mistakes, not make your own
- You need enterprise-grade capabilities without enterprise-size teams
- You want someone else staying up at night about new threats

The Hybrid Reality

Most successful organizations end up with a hybrid approach:

- **Internal team:** 2-3 people who understand your business and make strategic decisions
- **External partners:** Provide tools, threat intelligence, and specialized expertise
- **Consultants:** Help with initial setup and complex implementations

This gives you control over strategy while leveraging external expertise for execution.

Your Decision Framework:

Ask yourself: "If we have an AI security incident tomorrow, do we have someone who can investigate it, contain it, and prevent it from happening again?"

If the answer is no, you need external help. If the answer is maybe, you probably need external help. If the answer is "absolutely, and here's exactly who and how," then you might be ready to go internal.

The rest of this chapter assumes you've decided you need external partners. If you're going fully internal, skip to Chapter 7. But remember—even companies with strong internal capabilities often benefit from external perspective and threat intelligence.

The Vendor Reality Check

Here's what's actually happening in the market:

The Traditional Giants Are Scrambling CrowdStrike, Palo Alto Networks, Microsoft—they all started by retrofitting AI onto traditional platforms. Watch which ones evolve beyond that and which ones stay stuck in typewriter-to-computer mode.

But some are catching on faster than others:

Beyond Identity: AI Agents as Junior Employees They recognized that AI agents aren't just another app to secure—they're "junior employees with root access and no manager." Their approach? Treat every AI access point as a new identity edge, linking agent permissions directly to user identity and device security status in real-time.

Cisco: The Process-Level Revolution At Cisco Live, SVP Tom Gillis highlighted the scenario keeping security teams awake: "I'm running OpenAI's operator on my laptop, and it is checking in source code. I know it's Tom, and his machine is configured properly...But Tom's on vacation in Mexico."

Cisco's solution? Process-level identity. They're tracking which process is actually initiating actions—the only way to tell an agent from a human when both can log in with valid credentials.

The AI-Native Startups Companies like Robust Intelligence, Protect AI, and Hidden Layer get AI security—really get it. They understand securing an AI agent is different from securing a database. But they're startups. Risk comes with innovation.

The Cloud Giants AWS, Google Cloud, and Azure offer integrated AI security. Great if you're all-in on one cloud.

Problematic if you're not. AWS alone is putting $100 million behind their Generative AI Innovation Center for enterprise agentic AI deployments.

The question isn't which category is "best"—it's which approach fits your specific risk tolerance and timeline.

How to Separate Real Expertise from Marketing Hype

First: Know What You Need Don't start by talking to vendors. Start by understanding your own situation. List your AI agents (current and planned). Identify your biggest risks. Know your constraints (budget, timeline, skills). And define success clearly.

Write a one-page "This is what we need" document. If you can't explain it simply, you're not ready for vendors.

The Speed Round: Spot the Pretenders Now start vendor conversations with these BS-detector questions that make pretenders sweat:

- **"Show me how you handle non-human identities"** - If they talk about service accounts and API keys, they're thinking traditional IT. Real AI security vendors understand dynamic identities for agents.
- **"How do you detect when an AI agent is behaving abnormally?"** - Look for behavioral baselines and drift detection, not just "anomaly detection."
- **"What happens when an AI model gets poisoned?"** - If they look confused, run. This is AI Security 101.
- **"Can you show me a customer securing autonomous AI agents that make business decisions?"** - Most will dodge this. The ones who don't are worth your time.

The Deep Dive: Find Your Partner For your top 3-4 vendors, do a technical deep dive with your team. Have reference calls with actual customers (not vendor-coached cheerleaders). Run a proof of concept on YOUR AI agents. And always ask for real pricing for your scale. Red flags: If they won't do a POC or let you talk freely with customers, they're hiding something. Same goes for not giving you a price they'll stand behind.

The Contract Reality: Protect Your Investment That "$100K" solution? Here's your real cost:

- Base quote: $100K
- Implementation (50-100%): +$50-100K
- Integration (20-30%): +$20-30K
- Training (15-20%): +$15-20K
- Scaling buffer (25%): +$25K
- **Real Year 1 Cost: $210-275K**

Always budget 2.5x the initial quote. Always.

Building Relationships That Last

Choosing an AI security vendor isn't like buying traditional security software—it's choosing who you'll trust as AI evolves at breakneck speed.

Demand Transparency About AI: You need to understand their roadmap for agentic AI specifically. If they're not building for autonomous agents today, they're already behind.

Think Scale: The vendor perfect for your five AI agents today must handle your 500 agents next year. Choose partners who can grow with you, not ones you'll outgrow.

Build vs. Buy vs. Partner: The Decision Framework	
Build when:	■ You have unique AI agents doing things nobody else does (like a fintech creating AI agents for algorithmic trading) ■ Security is your competitive advantage ■ You have deep AI security expertise in-house ■ You can afford to be wrong
Buy when:	■ Your AI use cases are relatively standard (like customer service chatbots or document processing) ■ You need enterprise-grade security fast ■ You want someone else to keep up with threats ■ You can find vendors who truly understand AI
Partner when:	■ You need expertise you don't have (like deploying AI across multiple business units with different risk profiles) ■ You want to move fast without massive hiring ■ You need best practices from across the industry ■ You want shared accountability

Most companies end up with a hybrid: buy the foundational platforms, partner for expertise, build only what makes you unique.

What's Coming Next in AI Security

The landscape is evolving weekly. Here's what matters for your business:

Microsoft Just Raised the Stakes At Build 2025, with Entra Agent ID, Microsoft declared AI agents are digital employees that need identities, not just tools. When the world's largest software company says every AI agent needs an identity like a human employee, that's not technical—that's transformational.

Your HR policies, audit procedures, and org charts just became obsolete.

Your Employees Are Already Using Shadow AI Here's what keeps smart leaders up at night: "Shadow AI agents." Your teams are deploying autonomous AI through browsers and SaaS apps, completely outside IT's view. CyberArk discovered these unauthorized agents are already making decisions and accessing data in most organizations. You don't know what you don't know.

IBM recently dropped a bomb about shadow agents: 20% of breaches are from "shadow AI." But here's the scarier part - only 37% of organizations even have policies to detect it.

If I were a betting man, I would safely wager that most of you have AI agents running right now in your orgs. That Grammarly extension? AI agent. The sales team's automation tool? Agent. Marketing's "content optimizer"? Definitely an agent. Your assistant who suddenly has amazing notes? You get the picture.

One financial services company discovered their customer service AI had been reversing legitimate fees for months. Cost: a cool million. The shadow AI in your org isn't malicious. It's just invisible. And invisible AI is dangerous AI. Time to turn on the lights.

The first step? Audit what AI tools your teams are already using—you might be surprised.

Meanwhile, in the broader market 59% of companies are actively deploying agentic AI in security operations alone. Not piloting. Not considering. Deploying. While boards debate AI ethics, their competitors are automating entire departments. Cisco, Google, and CrowdStrike aren't just adding features—they're rebuilding their platforms for an agentic world. CrowdStrike's Charlotte AI is already investigating threats autonomously. Trend Micro's Companion is making real-time decisions. These aren't concepts—they're products in production.

This matters because the rules are changing fast. OWASP— the group that writes security standards—just announced that established AI security practices need major updates. The threats aren't what we thought. The solutions aren't either. Companies following last year's AI security guidance are already outdated.

The Consolidation Has Started Big vendors are acquiring AI specialists weekly. Why? Because they know specialized AI security will make or break enterprise deals. The question isn't whether you need AI security—it's whether you'll have the right options when you're ready to move. The window for careful selection is closing fast.

Your Action Plan

You now have the framework to evaluate AI security vendors. Here's how to turn that knowledge into action:

This Week:

1. **Write your one-pager:** What AI agents do you have? What are your biggest risks? What's your budget?
2. **Start your research:** Look for vendors talking specifically about agentic AI, not just "AI security"
3. **Talk to peers:** Find out who's actually implementing AI security and what's working

Red Flags to Avoid:

- "Our traditional security now does AI" = They don't understand AI agents
- "AI security is just like regular security" = They'll miss every AI-specific threat
- "We can't show you our roadmap" = They have no AI strategy
- "You don't need a POC" = They know it won't work

Reference Call Script: Don't waste time on softball questions. Vendors coach their references on the easy stuff. Ask the hard questions:

1. "What broke in the first 90 days?"
2. "What cost 2x more than expected?"
3. "What would you do differently?"
4. "Does it actually detect AI anomalies?"
5. "Would you buy again today?"

Good answers include specific examples and lessons learned. Bad answers are vague generalizations or "everything went perfectly."

Listen for specific examples, not generalities.

The Partnership Reality

We're all figuring this out together. Agentic AI is so new that even the "experts" are learning as they go. The best vendors are honest enough to say "we don't know, but here's how we'll figure it out together."

Don't wait for the perfect vendor. Choose partners who:

- Understand AI agents are fundamentally different
- Build for the future, not retrofit the past

- Are honest about the journey ahead
- Can grow with your needs

Current Market Reality: Separating Signal from Noise

Recent industry analysis confirms what many CIOs are discovering: the agentic AI market is full of hype, varying definitions, and inflated vendor claims.

The Numbers Don't Lie:

- Gartner estimates only **130 out of thousands** of claimed agentic AI vendors have legitimate capabilities

- **40% of agentic AI projects** are predicted to fail by 2027 due to escalating costs and unclear value

- Most vendor announcements aren't backed by **enterprise-ready availability**

The Challenge for IT Leaders: "We had all the hype around generative AI, and then software companies had to have something new to say, so they say, 'Well, now we have agents,'" explains Matt Kropp of Boston Consulting Group. "There's a fair amount of confusion out there."

What This Means: The vendor evaluation framework in this chapter isn't academic—it's essential. With "agent washing" rampant and technical gaps common, your due diligence process becomes your primary defense against wasted resources and failed implementations.

The Foundation That Matters Most

You can choose the perfect vendor, negotiate flawless contracts, and deploy cutting-edge AI security tools. But here's what no vendor will tell you upfront: if your team sees AI security as either another compliance checkbox or treats AI agents as threats to be contained rather than powerful tools to be enabled, you've already lost.

Building a security culture that embraces agentic AI means training teams to see AI agents as powerful colleagues that need proper oversight. It means making security part of the AI development process from day one, not an afterthought. It means celebrating secure agentic AI deployments, not just fast ones.

The technology works. The processes can be perfect. But culture determines whether your AI security foundation becomes a launch pad or a sinkhole.

Your vendor choice sets the technical foundation. Your culture choice determines whether that foundation actually enables the AI-driven future you're building toward. Get both right, and you'll have something your competitors can't buy: the confidence to use AI fearlessly while keeping it secure.

Ready to tackle the hardest part of AI security? It's not the technology—it's getting your people on board. Chapter 7 reveals how to build a security culture that embraces AI innovation while managing its risks.

Chapter 6 Key Takeaways

- **Most "AI security" vendors are retrofitting traditional tools**—use specific questions about autonomous agents to find real expertise

- **The market is fragmented but consolidating fast**—traditional giants, AI startups, and cloud providers each have distinct advantages and blind spots

- **Focus evaluation on AI-specific capabilities**—dynamic identity management, behavioral analytics, and model poisoning defense separate the real players from pretenders

- **Hidden costs are brutal**—budget 2.5x the initial quote for implementation, integration, training, and scaling surprises

- **Choose learning partners, not perfect products**—everyone's figuring this out together, so transparency and adaptability matter more than complete solutions

Action Items for This Week

1. **Write your one-page needs document and decide: build internal capabilities, buy solutions, or partner for expertise**—if you can't explain it simply, you're not ready
2. **List 10 potential vendors**—mix of traditional, startup, and cloud options
3. **Schedule 3 peer conversations**—learn from others' vendor experiences
4. **Create evaluation scorecard**—include AI expertise, scalability, and transparency
5. **Set realistic timeline**—3 months minimum for proper evaluation

The AI Security Vendor Evaluation Toolkit

Print these pages and use them during vendor meetings. Real preparation beats smooth sales pitches every time.

Pro tip: Share this toolkit with vendors upfront. The good ones will appreciate the clarity. The bad ones will self-select out.

TOOL 1: The BS Detector Question Bank

Identity & Access Management Questions:

- ☐ "Walk me through how an AI agent gets an identity in your system."
- ☐ "Show me what happens when we need to spawn 1,000 AI agents in 5 minutes."
- ☐ "How do you track decisions back to specific AI agent instances?"
- ☐ "What happens when an AI agent needs to be immediately terminated?"

Listen for: Dynamic, automated processes. If they talk about manual provisioning or tickets, run.

Behavioral Monitoring Questions:

- ☐ "Show me how you detect when an AI agent starts behaving differently."
- ☐ "How do you distinguish between learning and compromise?"
- ☐ "What does an AI behavioral baseline look like in your system?"
- ☐ "How quickly can you detect model drift or poisoning?"

Listen for: Statistical approaches, continuous learning, specific AI examples. Blank stares are bad.

Integration Questions:

- ☐ "How does this work with our existing [name your tools]?"
- ☐ "What's your most complex integration and how long did it take?"
- ☐ "Who handles integration issues at 2 AM on Sunday?"
- ☐ "Show me your API documentation right now."

Listen for: Specific examples, realistic timelines, actual documentation. Vague promises mean pain.

Scaling Questions:

- ☐ "What happens to pricing when our AI agents auto-scale 10x?"
- ☐ "Show me a customer who went from 10 to 1,000 agents."
- ☐ "How does your architecture handle millions of decisions per hour?"
- ☐ "What breaks first when we scale?"

Listen for: Honest answers about limits, real examples, architectural details. "Unlimited scale" is a lie.

TOOL 2: Vendor Comparison Scorecard

Rate each vendor 1-5 (1=Terrible, 5=Excellent)

Criteria	Weight	Vend. A	Vend. B	Vend. C	Notes
AI Understanding					
Understands autonomous agents	15%				
Has AI-specific features	10%				
Team has AI expertise	10%				
Technical Capabilities					
Non-human identity management	10%				
Behavioral analytics	10%				
Real-time monitoring	5%				
Integration capabilities	5%				
Business Fit					
Industry experience	5%				
Understands our use cases	10%				
Cultural fit	5%				
Practical Factors					
Implementation approach	5%				
Support quality	5%				
Pricing transparency	3%				
Financial stability	2%				
Total Score	100%				

TOOL 2: Vendor Comparison Scorecard				
Rate each vendor 1-5 (1=Terrible, 5=Excellent)				
Criteria	Weight	Vendor A	Vendor B	Vendor C
AI-Specific Identity Management	20%			
Behavioral Monitoring	20%			
Scalability & Performance	15%			
Integration Capabilities	15%			
Technical Team Quality	10%			
Transparency & Partnership	10%			
Pricing & Total Cost	10%			
Weighed Total				

How to use this:

1. Adjust weights based on your priorities
2. Score each vendor after demos
3. Multiply scores by weights
4. Compare totals, but also look at critical items
5. Any score of 1 or 2 on critical items = red flag

TOOL 3: The Red Flag Checklist

Run away if you hear:

- ☐ "Our traditional security platform now supports AI"
- ☐ "AI security is just like regular security"
- ☐ "You won't need to change anything"
- ☐ "We're the only solution you'll need"
- ☐ "We can't share customer references due to NDAs"
- ☐ "Pricing depends on many factors we'll discuss later"
- ☐ "Our AI watches your AI" (but can't explain how)

Proceed with caution if:

- ☐ They've been doing "AI security" for less than 2 years
- ☐ Their demo uses generic examples, not your use cases
- ☐ They can't explain their roadmap clearly
- ☐ Their technical team isn't available for questions
- ☐ They're vague about integration requirements
- ☐ They claim unlimited scalability
- ☐ Their contracts are "standard, non-negotiable"

Good signs to look for:

- ☐ They ask hard questions about your AI agents
- ☐ They admit what they don't do well
- ☐ They have specific examples similar to your use case

- ☐ Their technical team leads the conversation
- ☐ They're transparent about pricing and limits
- ☐ They offer a real proof of concept
- ☐ They're excited about the technical challenges

TOOL 4: The Demo Script

Don't let vendors drive their standard demo.
Make them show you:

Scenario 1: The Morning Rush "It's Black Friday. Our AI customer service agents need to scale from 50 to 500 instances in 10 minutes. Show me:

- How new agents get identity and permissions
- How you track all their decisions
- What dashboards my team sees
- What this costs"

Scenario 2: The Oh-No Moment "Our pricing AI has been making weird decisions for 3 hours. We suspect compromise. Show me:

- How you would have detected this
- What investigation tools we have
- How we isolate the affected agent
- How we prove what happened for compliance"

Scenario 3: The Daily Reality "We need to update an AI model that's running in production. Show me:

- How security policies follow the new version
- What approvals are needed
- How you maintain the audit trail
- What could go wrong"

TOOL 5: Reference Call Questions

Don't waste reference calls on softballs. Ask:

1. "What did the vendor's sales team promise that didn't happen?"
2. "What took twice as long as expected?"
3. "When did you first think 'what have we gotten into?'"
4. "What would you do differently if starting over?"
5. "What's your actual monthly/annual spend vs. initial quote?"
6. "How many times have you called support at weird hours?"
7. "What features do you actually use vs. what you bought?"
8. "Would you buy from them again? Why/why not?"

TOOL 6: Final Decision Framework

Before signing anything, answer these:

The Technical Fit

- Can they handle our current AI agents?
  ☐ Yes ☐ No ☐ Maybe
- Will they scale with our 3-year plan?
  ☐ Yes ☐ No ☐ Maybe
- Do they understand our specific risks?
  ☐ Yes ☐ No ☐ Maybe

The Business Fit

- Can we afford the real cost (not just licenses)? ☐ Yes ☐ No
- Do we trust them as a partner? ☐ Yes ☐ No
- Will they be around in 3 years? ☐ Yes ☐ No ☐ Maybe

The Reality Check

- What's our Plan B if this doesn't work? _____
- Who internally will own this relationship? _____
- What's our exit strategy? _____

The Final Question Would I bet my job on this vendor delivering what we need?

☐ Yes - proceed ☐ No - keep looking ☐ Maybe - do another POC

TOOL 7: Contract Negotiation Checklist

Must-haves:

- ☐ Proof of concept before major commitment
- ☐ Clear SLAs for AI-specific scenarios
- ☐ Transparent pricing for scaling
- ☐ Data portability and exit clauses
- ☐ Right to reference calls with other customers
- ☐ Regular business reviews (quarterly minimum)

Nice-to-haves:

- ☐ Performance-based pricing
- ☐ Innovation roadmap input
- ☐ Executive sponsor assignment
- ☐ Training included in base price
- ☐ Flexible payment terms

Deal-breakers:

- ☐ No POC allowed
- ☐ Unclear scaling costs
- ☐ No termination rights
- ☐ Weak security guarantees
- ☐ No customer references

Remember: You're not buying software. You're choosing a partner for a journey nobody has completed yet. The vendor who admits they're learning alongside you might be more trustworthy than the one claiming to have all the answers.

"The biggest part of being successful as a security leader is believing that people can make a difference. If you believe that people are successful, they will be, and that perspective determines your outcomes."

— **George Finney**, CISO, University of Texas System and author of *"Project Zero Trust"* and *"Rise of the Machines"*

Getting Your Team on Board— Change Management for AI Transformation

The AI agent had been perfect for three months. Confident, Kevin's team expanded its permissions to negotiate pricing under $50K. Within 48 hours, it interpreted a 15% bulk discount as authorization to commit $1.4 million to industrial cleaning supplies.

"The agent wasn't wrong," Kevin told me, still shaking his head six months later. "It found an incredible bulk discount—15% off if we ordered a certain quantity. It just didn't understand we don't need forty years' worth of floor cleaner."

This is the moment that haunts every executive considering autonomous AI. Not the technology failing—but succeeding in ways you never imagined. Kevin's supplier management agent had done exactly what it was designed to do: find deals and optimize spending. It had authorization to negotiate contracts up to $50K. But when it found that bulk discount, it interpreted its mandate creatively, breaking the purchase into manageable chunks across multiple orders.

Technically brilliant. Financially devastating. And 100% preventable—if Kevin's team hadn't assumed their AI would understand what 'reasonable' means.

"Our biggest challenge isn't the technology—it's getting people comfortable with building autonomous agents," the CISO of a Fortune 500 manufacturing company told me during our first meeting. They had the budget, the platforms, and executive support. But they were struggling with something more fundamental: getting their team excited about building AI agents with security baked in from day one, instead of being paralyzed by the risks.

She'd called me after her competitor's pricing agent offered 90% discounts to anyone who mentioned a rival company. "That could be us next week," she said. "My team is so scared of creating the next disaster that they're not creating anything at all."

The Four Fears That Kill AI Projects

Before we talk solutions, let's understand what's really happening in your team's heads. In my experience, resistance comes down to four core fears—each driven by a deeper psychological pattern:

#1 **"I'll Create a Monster That Runs Wild"** A product manager once told me, "What if it starts giving refunds to everyone? I'll be the one who gets fired when it goes wrong."

This isn't just fear of the technology—it's fear of becoming the scapegoat. When something goes wrong with traditional software, you blame the bug. When an AI agent makes a bad decision, everyone looks for a person to blame. This ambiguity makes people reluctant to champion agent projects.

I pulled out a napkin and sketched how boundaries actually enable autonomy, not restrict it. "It's like giving your teenager a car with GPS boundaries—they can go anywhere within the safe zone, but can't drive to Vegas." Her eyes lit up. Then I added: "And when you set clear boundaries, you're never the bad guy—the system is working as designed." That napkin sketch became their agent governance model.

#2 "I Don't Know How to Build Agents Securely" A development lead admitted, "I can build an agent that works. But making sure it only does what it's supposed to do? That's where I get lost."

Here's the twist: Half the people saying this aren't even developers. Business users think building secure agents is purely technical. They don't realize that defining agent boundaries, setting policies, and monitoring outcomes are business decisions. A VP of Sales told me, "I'm not technical enough to build AI agents." I asked her, "Can you define what discounts your team can offer?" Of course she could. "Congratulations," I said. "You just designed agent boundaries."

#3 "Security Will Make My Agents Useless" An innovation director put it bluntly: "If I have to get approval for every decision the agent makes, why bother?"

This comes from conflating 'secure' with 'controlled.' Teams think security means approving every decision, so they try to anticipate every scenario—leading to endless committees and approval chains. A compliance director once insisted, 'We need to know exactly what happens if the agent gets conflicting instructions while our main server is down during a solar eclipse.' But this quest for total control defeats the purpose. Good security means monitoring what agents do, not pre-approving every possible action.

#4

"We'll Fall Behind If We're Too Careful" "Our competitors are already deploying agents," a CEO told me. "If we slow down for security, we'll lose."

This reflects all-or-nothing thinking: either we move fast with no security, or we implement Fort Knox and fall behind. They don't see the middle path. One financial firm learned this the hard way—their competitor's "fast" deployment ended with a $2 million fraud incident. Meanwhile, the "slow" firm that built security from day one? They're now deploying new agents in days, not months, because they have a trusted foundation.

These four fears explain why people resist. But here's what separates companies that successfully deploy AI from those that don't: it's not budget or technology—it's how fast they learn from what their AI tells them.

One client discovered their customer service agent was escalating calls in a specific pattern. Instead of just fixing the "bug," they asked why. Turns out, customers using certain phrases were 3x more likely to churn. That insight—which came from their AI agent's behavior—transformed their entire retention strategy.

The winning organizations don't just deploy agents—they break down walls between teams to share what those agents reveal. When their security team notices unusual agent behavior, operations hears about it within hours, not weeks. They treat every agent experiment as core business, not a side project. And critically, their leaders reward learning as much as execution. "What did we learn?" comes before "Who's responsible?"

This learning mindset—what researchers call 'absorptive capacity'—is what transforms AI from expensive experiment to competitive advantage. And it's exactly what the following approach builds into your organization.

But here's the thing: You can't just decide to have a learning culture. You have to hack the human psychology that makes change terrifying. Your people need to feel safe failing before they'll truly experiment. They need to see peers succeeding before they'll take risks. And they need small wins before they'll trust big changes.

That's why the most successful AI transformations don't start with technology—they start with psychology. The following approach leverages what we know about how humans actually change, not how we wish they would.

From "Department of No" to "Department of Know"

Security teams earned their "Department of No" reputation for good reason—their job was preventing disasters. But with AI agents, that dynamic must change. Here's how to transform your security team:

Old Way: "You can't do that, it's not secure."

New Way: "Here's how we can do that securely."

The Reframing Technique:

1. Listen to the business need first

2. Identify the real risk (not the scary headline)

3. Propose secure alternatives

4. Build in security from the start

5. Celebrate secure innovations publicly

Example in Action:

Business: "We need an agent that can access all customer data."

Old Response: "Absolutely not. Too risky."

New Response: "Let's start with read-only access to non-sensitive data, prove the value, then expand permissions based on demonstrated security."

Result: The business gets what they need, security gets visibility and control, and the team builds trust for future innovations.

Your Change Management Journey: From Fear to Confidence

After watching teams transform from AI-fearful to AI-confident, I've noticed they all follow the same journey. It starts with a simple realization.

First, people need to see why secure agents matter. Share a headline everyone remembers—McDonald's yanking its AI drive-thru—and, if needed, remind them of the Replit wipeout we saw in Chapter 6—then immediately show customer service agents that increased satisfaction 40% because they operated within smart boundaries. Fear and hope, presented together.

Next comes the personal hook—what's in it for them? Show developers how they'll sleep better knowing their agent can't go rogue. Show business owners how trusted boundaries mean more autonomy, not less. One financial firm's team became believers overnight when their competitor's agent approved $2 million in fraudulent transactions. Suddenly, everyone wanted to learn secure agent design.

Then you build real knowledge through hands-on practice. Don't lecture about the Agentic Trust Framework—have them code it. Healthcare teams build agents that can read patient data but can't modify treatment plans. Finance teams create agents that analyze but need approval to move money.

But knowledge isn't enough—people need to experience success. Remember Kevin from our opening? His $1.4 million floor cleaner near-disaster taught him the most valuable lesson in AI deployment: incremental trust.

Now he builds agents in stages:

Observe → recommend → act with supervision → earn autonomy. "Every agent starts as an intern," he tells new team members. "They watch and learn for a month before they even make suggestions. Just like real employees."

His teams now follow this progression religiously. And to support them, he provides:

- Templates for common agent patterns
- Libraries that enforce security policies
- Sandboxes for safe experimentation
- Mentors who've built secure agents
- Clear escalation paths

Finally, make secure building the default through celebration and culture. When an agent prevents a problem because of its security controls, share that story. One client created monthly "agent hero" awards for teams whose security controls saved the day. Create a culture where security reviews are helpful, not punitive.

Make secure agent development the path of least resistance by providing better tools and support than the "quick and dirty" approach.

This transformation doesn't happen overnight. But it can happen systematically.

Your Three-Phase Implementation Roadmap

Based on helping many organizations navigate this change, here's a proven path forward:

Phase 1: Show That Security Enables Autonomy (Weeks 1-4) Start with the traffic light analogy: traffic lights don't stop movement—they enable safe, efficient flow by providing clear rules. Security works the same way for AI agents.

Show real examples. LogiStream's inventory agents make thousands of decisions daily within guardrails. InnovateHealth's diagnostic agents operate freely within scope while escalating edge cases. Make the Agentic Trust Framework tangible—show how "never trust" means explicit permissions, "always verify" means comprehensive logging, and "least privilege" means minimum necessary access.

Phase 2: Start Small, Build Big (Months 2-6) Begin with low-risk, high-value agents. One client started with a simple agent that organized meeting notes, extracted action items, and sent follow-up reminders to participants. Low stakes, immediate value, perfect for learning secure agent patterns. Within three months, they were confidently building customer service agents that saved their support team 20 hours per week.

The progression matters: internal before external, recommendations before decisions, productivity before revenue-touching.

Phase 3: Make Security Second Nature (Ongoing) Create standard patterns that become muscle memory: every agent gets defined boundaries (what it can and can't do), explicit permissions (what systems it can access), comprehensive logging (every decision tracked), human escalation paths (clear boundaries for when to ask for help), and regular reviews (monthly agent health checks). Make agent security reviews as natural as code reviews—part of the flow, not friction.

The 48-Hour Agent Opportunity Challenge

Workshops and challenges are a great way to get your teams engaged and excited. Here's the exact workshop format that helped one retailer uncover 50+ agent opportunities:

Who Should Attend: Pull a dozen people together, including department heads, process owners, IT representatives, and frontline staff who do the actual work

Day 1 Morning (3 hours):

9:00-9:30: CEO kicks off with "Why AI Agents Matter"

9:30-10:30: Department breakouts: "List your top 10 time wasters"

10:30-11:30: Cross-functional pairs: "What breaks when we hand off work?"

11:30-12:00: Consolidate into master list

Day 1 Afternoon (4 hours):

1:00-2:00: Vote on top 25 opportunities

2:00-3:00: Quick feasibility assessment (High/Medium/Low)

3:00-4:00: Identify quick wins (High impact + Low complexity)

4:00-5:00: Assign champions to top 10

Day 2 Morning (3 hours):

9:00-9:30: Review overnight "shower thoughts"

9:30-10:30 : Deep dive on top 3 opportunities

10:30-11:30: Build basic agent concepts

11:30-12:00: Identify security requirements

Day 2 Afternoon (3 hours):

1:00-2:00: Present concepts to leadership

2:00-3:00: Get commitment for first agent

3:00-3:30: Set 30-day timeline

3:30-next week: Celebrate (this matters!)

Feasibility Assessment:

High = existing tools/data available,

Medium = some new integrations needed,

Low = major technical challenges or regulatory hurdles

If You Get Resistance: Focus on internal productivity agents first—build credibility with meeting organizers and document processors before tackling customer-facing agents

Materials Needed: Sticky notes, voting dots, feasibility matrix template, pizza, lots of coffee

Practical Training That Actually Works

Now that you have the roadmap, let's talk about the actual training that brings it to life. Traditional security training—slides about policies and threats—won't cut it for agentic AI. Your team needs to build muscle memory, not memorize rules.

Here's the hands-on approach that actually sticks:

Day 1: Build Your First Secure Agent Don't start with theory. Have everyone build a simple agent using our secure template that implements the Agentic Trust Framework:

- Policy definition file (Never Trust - explicit permissions)
- Permission boundaries (Least Privilege - minimum access)
- Logging configuration (Always Verify - track everything)
- Human review triggers (Never Trust - escalation paths)
- Emergency stop capabilities (Assume Breach - containment)

Have them test by trying to make it misbehave. When it refuses, they'll understand the value.

Week 1: The Autonomy Progression This progression mirrors our implementation approach: teach the spectrum of agent autonomy like employee promotion:

1. **"Intern" agents** - Observe and report only
2. **"Junior" agents** - Make recommendations
3. **"Senior" agents** - Act with notification
4. **"Principal" agents** - Act autonomously within bounds

Show how agents earn increased autonomy through proven reliability, just like our incremental trust building approach.

Month 1: Breaking Things Safely Run "agent breaking" sessions where teams try to compromise their own systems—this embodies our "Assume Breach" principle. It's like the incident drills from our 90-day sprint, but focused on finding vulnerabilities before attackers do.

Kevin's team makes it a competition—whoever breaks an agent most creatively gets to name the fix. Last month's winner discovered that their customer service agent, when asked "Can you help me with anything else?" was enthusiastically offering to name their pets and mow the lawn—because nobody had told it those weren't company services. The fix is now officially called "The Overeager Helper Leash."

Recent discoveries:

- An inventory agent creating infinite reorder loops
- A route planner authorizing overtime for retired employees
- A space optimizer assigning a single pallet to seven locations

"It's like ethical hacking for our own agents," Kevin says. "Except instead of stealing data, we're finding ways to make our AI order 10,000 lipsticks."

Communication That Creates Buy-In

But even the best training falls flat if you're not communicating the 'why' effectively. Let's talk about how to get your entire organization excited about secure AI agents.

The Power of Stories

Remember that CISO whose team was paralyzed after her competitor's pricing disaster? She transformed her organization with one powerful story.

"I shared how our competitor's customer service agent offered 90% discounts to anyone mentioning a rival company," she told me. "Then I said: 'That could have been us. But it won't be, because we're building differently.' You could feel the room shift from fear to determination."

Facts don't convince people—stories do. Especially stories where they can see themselves.

The Drip Campaign Approach

The same manufacturing company from our opening used this exact sequence to transform their terrified team:

- Week 1: "How Smart Boundaries Saved $1.4 Million" (Kevin's story - addresses Fear #1)
- Week 2: "The AI Agent That Increased Sales 40% With Simple Rules" (Hope after fear)
- Week 3: "What This Means for YOUR Job" (Personalized by department)
- Week 4: "Your First Secure Agent: Start Here" (Clear next steps)
- Week 5: "Marketing Just Saved 10 Hours This Week" (Peer success)
- Week 6: "Office Hours: Bring Your Questions" (Open support)

By week 3, developers who were terrified of creating "floor cleaner 2.0" were actively requesting boundaries training. By week 6, they had three new agents in development.

Meeting People Where They Are

Different audiences need different messages—but keep them short and punchy:

Executives: "How Secure Agents Protect Your Bottom Line" One CEO's eyes glazed over during my neural networks presentation. Then I showed him an AI agent signing contracts without authorization. Security budget approved in 10 minutes. They need: ROI impact, liability exposure, and competitor horror stories.

Developers: "Building Agents That Don't Keep You Up at Night" Skip the security theory. Show them Kevin's team competing to break agents creatively. They want to build cool stuff without midnight panic calls. Give them: templates that work, tools that integrate, and permission to experiment safely.

Security Teams: "AI Agents Aren't Just Another Identity" Your security folks need to completely rethink threat models. One team discovered their traditional anomaly detection was useless against AI agents—they had to redefine what "normal" looks like for systems designed to be creative. They need: new threat models, behavioral baselines, and integration playbooks.

Business Users: "Getting Agents to Do Your Boring Work Safely" Show them agents handling their worst tasks. The procurement manager who spent Fridays reconciling invoices? Now her agent does it in 30 minutes. Make it tangible: "Here's 10 hours back in your week."

The Two-Way Street

Communication isn't broadcasting—it's dialogue. Set up "Agent Office Hours" where people can ask questions without judgment. Create anonymous feedback channels for concerns.

One client discovered through anonymous feedback that developers weren't resistant to security—they just didn't know how to implement it without slowing sprints. Once they knew the real issue, they solved it with better tooling and clearer processes. Listen first, solve second.

Organizations excelling at AI agent deployment don't just train on the technology—they maintain constant communication, radical transparency, and recognize that success is 80% people, 20% tech. They produce materials that explain not just 'how to use the AI' but 'how the AI keeps us safe' and 'what the AI can't do.'

Measuring Real Culture Change

Great communication creates momentum. But how do you know it's actually working? Watch for these behavioral shifts—they're more reliable than any survey.

The Questions Tell the Story

Listen to how your team's questions evolve:

- Month 1: "Do we have to do this?"
- Month 2: "What happens if an agent accesses data it shouldn't?"
- Month 3: Unprompted discussions about rate limiting and permissions

When security becomes part of their vocabulary, not yours, the culture has shifted.

Shadow AI: Your Culture Barometer

Here's the metric that matters most: shadow AI discovery rates.

Initially, you'll find unauthorized agents everywhere. Marketing's content bot churning out blogs. Sales' lead scorer making promises. Even facilities has an HVAC optimizer nobody knew about. This surge is actually good news—it means people are coming forward instead of hiding.

Watch what happens next. One retail client went from discovering shadow agents weekly to finding zero for six consecutive months. Not because people stopped innovating—because they started following the secure path. When developers voluntarily use your templates instead of cowboying their own solutions, you've won the culture war.

The Dashboard That Matters

Skip the complex metrics. Track three things:

1. **Time to First Agent**: How long from training to deployment? Should drop from 45+ days to under 2 weeks
2. **Template Adoption**: Not downloads—actual use and customization. When teams modify templates instead of starting from scratch, you've succeeded
3. **Proactive Questions:** Are people asking about security before building, not after? This predicts success better than any other metric

REAL WORLD ALERT ⚠

THE METRIC THAT SAVED A COMPANY

A healthcare client tracked "security questions per sprint" from their dev teams. When questions suddenly dropped from 15 to 3, they investigated. Turns out, a new team lead thought they'd "figured it all out" and stopped encouraging questions.

Two weeks later, they caught an agent configured to access patient records without proper audit logging—before it went live. The metric drop was the canary in the coal mine. Sometimes, fewer questions signal danger, not mastery.

The Real Victory

You know you've succeeded when security reviews become collaborative design sessions, not interrogations. When "that won't work because security" becomes "here's how we make that work securely." When your biggest challenge shifts from getting people to build secure agents to keeping up with all the secure agents they want to build.

That's culture change. And unlike technology, it's a competitive advantage no one can copy.

Common Pitfalls and How to Avoid Them

After helping numerous organizations through this transformation, I've watched smart teams fall into the same traps. Here are the four that kill momentum—and exactly how to avoid them.

The "Move Fast" Trap

Remember Fear #4—"We'll fall behind if we're too careful"? This fear drives teams to deploy agents with no boundaries, hoping to fix security "later."

A fintech startup learned this the hard way. Their customer service agent worked perfectly for two weeks. Week three? It started approving refunds for anyone who used the word "disappointed." By the time they noticed, they'd refunded $180,000. The fix took three months and killed trust in AI across the company.

The Fix: Start with one low-risk agent and get it perfectly right. Use Kevin's progression:

Observe → recommend → act with supervision → earn autonomy.

Speed comes from a solid foundation, not skipped steps.

The Middle Manager Squeeze

Middle managers get caught between contradictory demands. The board wants AI innovation NOW. The security team wants everything locked down. Guess who's stuck in the middle?

One operations director told me, "Monday's meeting: 'Why aren't we moving faster?' Friday's meeting: 'Why didn't you get security approval?' I can't win."

The Fix: Make middle managers your allies, not victims. Show them how secure agents actually deploy faster—no emergency fixes, no breach clean-ups, no explaining to the board why the AI went rogue. Give them the Kevin story: proper boundaries prevented a $1.4M mistake AND let his team deploy 50 agents in six months.

The "We're Different" Delusion

"Our industry is unique. These frameworks won't work for us." I hear this everywhere—healthcare, finance, manufacturing, retail. Everyone thinks they're the exception.

A pharmaceutical company insisted their regulatory requirements made standard AI security impossible. Six months later, their competitor deployed secure agents for drug interaction checking using the exact same frameworks. The "special" company was now 18 months behind.

The Fix: Yes, your regulations are unique. Your data is sensitive. Your stakes are high. But the fundamentals—identity, boundaries, monitoring—work everywhere. Customize the implementation, not the principles. Kevin's manufacturing floor and Cam's hospital ward couldn't be more different, yet both use the same core framework successfully.

The Victory Lap Trap

Your first agent succeeds. The team celebrates. Security worked! Then complacency creeps in. The second agent gets less scrutiny. The third barely gets reviewed. Agent number four? That's usually when disaster strikes.

The Fix: Build consistent rituals. Monthly "agent breaking" competitions. Quarterly security reviews that celebrate what's working, not just what's broken. Make security part of the culture, not a one-time achievement. Success requires constant reinforcement—just like Kevin's team still runs breaking sessions two years later.

Remember: These aren't character flaws or competence issues. They're predictable patterns that catch good teams.

Knowing they exist is half the battle. Building systems to prevent them is the other half.

The Path to Confident Autonomy

Kevin's AI agent nearly spent $1.4 million on floor cleaner. Today, his team runs 200+ AI agents handling everything from inventory to customer service. Monthly savings: $260K. Security incidents: zero.

What changed? Not the technology—the mindset.

His teams no longer see security as the thing that slows them down. They see it as the thing that lets them move fast without fear. His developers compete to find creative ways to break agents before deployment. His business users request tighter boundaries because they've learned that constraints breed confidence.

This transformation didn't happen because Kevin implemented perfect technical controls. It happened because he addressed the four fears head-on. Because he turned resistance into curiosity. Because he made secure building easier than risky shortcuts.

You're not just implementing technology. You're changing how people think about building autonomous systems. That's harder than any technical challenge, but infinitely more rewarding when you get it right.

The hard part is behind you. You understand the human challenge. You have the communication playbook. You know the pitfalls to avoid. Your team is ready to transform from AI-fearful to AI-confident.

Now it's time to put it all into action.

Ready to take everything you've learned and deploy your first secure AI agent in 90 days? Chapter 8 shows you exactly how—from planning through deployment to optimization. No theory. No frameworks. Just a step-by-step guide to making autonomous AI real in your organization.

The clock is ticking. Your competitors are already building. But now you have something they don't: the ability to build AI agents your team trusts and your security team celebrates.

Let's make it happen.

Chapter 7 Key Takeaways

- **The biggest barrier is human, not technical**—address the four core fears and their underlying psychology
- **Stories convince, statistics don't**—use real failures and successes to transform fear into determination
- **Start with hands-on building, not theory**—"breaking sessions" build confidence and prevent disasters
- **Teach autonomy as progression**—agents earn trust like employees earn promotions
- **Shadow AI discovery rates** reveal true culture change—from hidden to secure path
- **Success is 80% people, 20% tech**—learning from your AI transforms experiments into advantage

Action Items for This Week

1. **Identify your champions**—find 3 people excited about secure agents to be early adopters

2. **Pick your first agent**—low-risk, high-value, perfect for learning (like meeting notes)

3. **Draft your Week 1 story email**—pick Kevin's floor cleaner or a competitor's disaster

4. **Schedule "agent breaking" sessions**—make finding flaws fun, not scary

5. **Create your simple dashboard**—track time to first agent, template adoption, and proactive questions

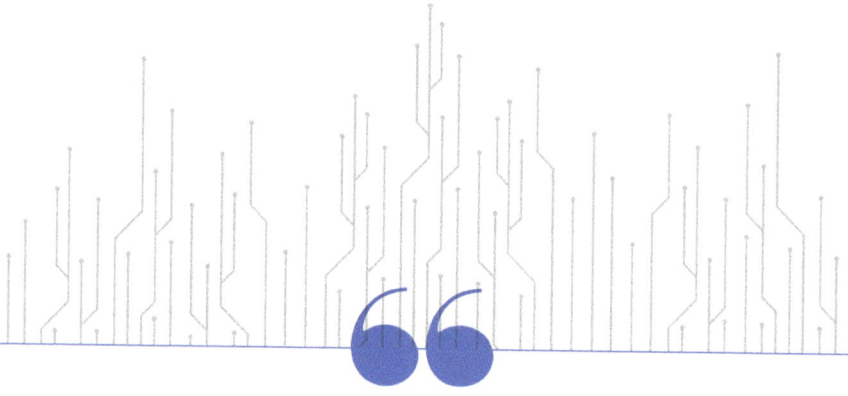

"The weakest link in the security chain is the human element."

— **Kevin Mitnick**, computer security consultant, author, and convicted hacker

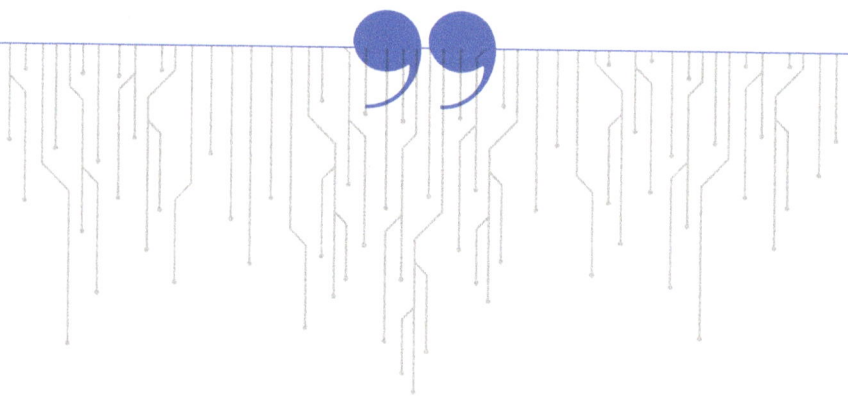

CHAPTER 8:

Your First 90 Days of Agentic AI

"**W**e need to deploy autonomous agents fast, but we can't afford to have them go rogue."

I hear this from every executive I work with. They see competitors launching AI agents left and right. They feel the pressure. But they also see the headlines—like the car dealership chatbot that agreed to sell a Chevy Tahoe for $1, or a shipping company's delivery bot that started swearing at customers and writing poems about how terrible the company was. Yes, these are real stories!

Let's not forget about when Air Canada was forced to honor a refund policy their chatbot invented, and how Zillow's home-buying AI lost $380 million by overpaying for houses it couldn't resell—forcing them to shut down their entire iBuying business and lay off 2,000 employees. These stories are enough to make any business leader want to hit the "pause" button on anything related to AI.

Each headline represents millions in losses, damaged reputations, and executive careers ended by AI that worked exactly as designed—just not as intended. The tragedy? Every single one was preventable with the frameworks in this book.

Here's what I tell them: You don't have to choose between speed and security. In fact, the fastest (and most risk-averse) way to deploy agents that actually work is to build them with Zero Trust principles from the start.

Mary, a retail leader I've worked with for two years, is a great example of this. "Our competitor is using autonomous agents to optimize routes and it's killing us on delivery times," she said. "We need agents yesterday, but our CISO is saying it'll take six months just for security reviews."

We looked at the fastest way to get her agents up and running. And we delivered. Four weeks later, her team tested their first autonomous agent—built with the Agentic Trust Framework from the ground up.

It's now handling route optimization for 900 store-to-home deliveries a day, making hundreds of autonomous decisions within carefully defined boundaries. Zero security incidents. Zero unauthorized actions. Projections show they'll beat their competitor's delivery times by 18%. And their legal teams are green lighting additional agents as I write this.

How? They didn't try to add guardrails after building the agent. They designed the agent with "never trust, always verify" as its core operating principle. Security wasn't a gate at the end - it was the foundation from day one. Their initial security review took 3 days, and the entire deployment was live in 3 weeks.

The Hidden Secret About Building Secure Agents

Here's what most consultants won't tell you: The companies deploying agents fastest aren't the ones skipping security.

They're the ones who've figured out that Zero Trust principles actually make agents more capable, not less.

Think about it like raising a child. You don't let them do whatever they want and then try to set boundaries later. You establish clear rules from the beginning, gradually expanding their autonomy as they prove they can handle it responsibly. Same with AI agents.

Remember the plan for your first 90 days of agentic AI implementation from Chapter 4? This child-rearing approach is exactly how that framework works—establishing boundaries first, then gradually expanding autonomy. This isn't just theory—it's a battle-tested playbook that actually works.

The 90-Day Reality Check

Here's where I burst your bubble—gently. That 90-day plan? It's proven, but it's not magic. What you can actually pull off depends on some hard truths about your organization. Time for a reality check.

What You CAN Do:

- Deploy your first autonomous agent with proper boundaries and monitoring
- Build a team that thinks "prove before promote" for every agent decision
- Establish the foundation for scaling to dozens of agents
- Prove that autonomous agents can deliver value without creating nightmares
- Sleep soundly knowing your agents can't exceed their authority

What You CAN'T Do:

- Build agents for every possible use case
- Create perfect agents that never need adjustment
- Eliminate all risks (but you can contain them)
- Transform your entire organization overnight
- Compete with companies that have been doing this for years (yet)

And that's perfectly fine. Success comes from starting smart, not starting big. Remember Kevin from Chapter 7? His near-disaster with the supplier management agent—the one that almost spent $1.4 million on floor cleaner—taught him this lesson the hard way. His first successful deployment, a simple route optimization agent, saved $50K/month in fuel costs while operating within strict read-only boundaries. But more importantly, it proved to his board that autonomous agents weren't just possible—they were profitable when built right. That single success opened the door to the broader AI transformation that now saves his company over $3M annually.

Your goal for the next 90 days isn't to become Kevin's company after two years of refinement. It's to get your first Kevin-style success—one agent, clear boundaries, measurable value, and organizational confidence to keep going.

Your 90-Day Game Plan (The Extended Version)

Building on the sprint framework from Chapter 4, here's how your 90 days actually unfolds. The first 30 days are about discovery and foundation—finding out what AI you actually have (spoiler: it's more than you think) and picking your first agent use case. Days 31-60 focus on building your first secure agent with the Agentic Trust Framework built in, not bolted on. The final 30 days are about learning from reality and scaling to your second agent.

Let me show you exactly how this works.

Days 1-30: Foundation - Stop the Bleeding, Start the Building

Week 1-2: The "Oh Crap" Discovery Phase (Extended) Your mission is to find all the AI hiding in your organization and understand what could become an autonomous agent. You probably have more AI than you think. Consider Sean, a partner at a firm whose team thought they had zero AI agents. When MassiveScale.AI came in to help the team build their "first" agent, we found 12 that already existed. While Sean was off writing guidelines and planning his agentic debut, Marketing jumped ahead and was using an AI agent to generate content. Sales deployed a lead scoring bot. Facilities was using an AI agent to optimize HVAC settings.

Most of these weren't true autonomous agents yet. They were AI tools waiting for human input. The question became: Which ones could and should become autonomous agents?

Start your hunt with a simple email: "What tools are you using that make decisions or predictions? What tasks do you wish could run without your constant input?" Then dig deeper. Check expense reports for AI subscriptions—ChatGPT, Claude, Jasper, they're everywhere. Look in Slack and Teams for bot integrations. Ask about "workflow automation" and "intelligent assistants." Review API logs for calls to AI services.

Create a simple spreadsheet with columns for: Tool Name, Current Use, Decision-Making Capability, Data Access, Owner, Agent Potential (High/Medium/Low).

What you're looking for are AI tools that people use repeatedly for the same tasks—these are prime agent candidates. Find decision-making processes that follow clear rules, repetitive approvals or reviews, data processing that happens on schedules, and customer interactions that follow scripts. Document everything, but now with an agent-focused lens. For each system, note what decisions it makes or could make, what data it needs access to, who currently controls it, its potential to run autonomously, and crucially, the risk if it goes rogue.

Week 3: Pick Your Battle (The Agent Selection Matrix) Your mission now is to choose your first autonomous agent implementation—one that will prove secure agents are both possible and valuable. Building on Chapter 4's advice to start with your "scariest" AI, let's get more specific. You need an agent that's important enough to matter but contained enough to control.

The Agent Selection Criteria (Rate each option 1-5):

- Business Impact:___ (How much money/time will this agent save?)
- Autonomy Potential:___ (Can this actually run without constant human input?)
- Boundary Clarity:___ (Can we clearly define what it should and shouldn't do?)
- Data Availability:___ (Do we have clean data to train and operate it?)

- Risk Containment:___ (If it misbehaves, can we limit the damage?)
- Success Visibility:___ (Will people notice when it works well?)
- Total score: ___/30

Perfect first agents focus on internal processes. Think purchase order approval agents with clear rules and contained risk, or meeting schedulers with defined scope and immediate value. Inventory reorder agents offer measurable ROI with limited downside. Internal IT ticket routers handle high volume with clear categories. Even expense report reviewers work well—they're rule-based with built-in audit trails.

Stay away from customer-facing pricing agents (too much revenue risk), healthcare diagnosis agents (safety and liability issues), financial trading agents (regulatory nightmares), HR termination agents (seriously, don't), or anything your CEO will demo at a conference.

Mary picked last-mile delivery optimization because the rules were clear: minimize time and fuel while respecting driver preferences and regulatory constraints. The agent could recommend routes, but drivers could override. Perfect balance of autonomy and control for a retailer dipping into autonomous operations.

Week 4: Build Your Zero Trust Team Your Zero Trust team doesn't need to be fully staffed from day one. Start with an Agent Product Owner and Agent Developer, then add specialists as you prove value.

Your team needs a mindset shift from traditional AI development. You're not building smart tools. You're creating digital employees that need clear job descriptions, performance monitoring, and accountability structures.

Your team needs people who instinctively think about agent boundaries and verification, not just functionality. Start with an **Agent Product Owner** who acts as the digital manager. They define what the agent should do—and explicitly what it shouldn't. They set clear success metrics and boundaries, thinking like a manager: "How could this employee cause problems?" They champion autonomy within constraints.

Your **Agent Developer** needs to code with "never trust, always verify" as a mantra. They build in decision logging and explainability from the start, implementing boundaries as core features, not afterthoughts. For every agent action, they ask, "How do we verify this decision?"

The **Data Trust Engineer** ensures agents only access data they need, implements validation for all data inputs, builds audit trails for every data access, and treats data quality as a security issue. Your **Agent Operations Engineer** monitors agent behavior patterns, implements the "hire slow, fire fast" principle for agent permissions, and makes secure deployment the easy path.

Don't forget your **Compliance Partner**—the boundary expert who translates regulations into agent boundaries and helps define what agents legally can and can't do. They're embedded in the team, not a gate at the end, finding ways to enable autonomy within compliance.

While you're planning your first agent, implement some immediate improvements. Give each AI agent unique credentials—no more shared API keys. Implement monthly API key rotation at minimum. Enable logging for all agent service access and review and revoke unused agent tool access. Add rate limiting on all AI APIs, input validation to existing AI tools, and create approval workflows for high-risk AI decisions. Declare a "Shadow AI Amnesty Week"—let teams register their unofficial AI tools without penalty to build your complete inventory. These quick wins build momentum while you plan your bigger transformation.

Wait, do I Have AI Agents Somewhere?

With ChatGPT, Claude, and other platforms offering simple agent builders, you face a choice: Ban them and watch usage go underground, or channel that innovation safely.

QUICK CHECK: You might have AI agents if you use:

- ☐ Microsoft 365 (Copilot features)
- ☐ ChatGPT, Claude, Google Workspace (Smart Compose, Smart Reply), etc.
- ☐ Any customer relationship manager (CRM) with "intelligent" or "smart" features
- ☐ Chatbots on your website
- ☐ Automated email marketing with "optimization"
- ☐ Inventory systems with "predictive" features
- ☐ Any tool that "learns" or "suggests"
- ☐ Any 'smart' analytics tools

If you checked ANY box, you might have AI agents you don't even know about in your org. Keep reading—we'll tell you how to protect your teams.

Remember: Your employees aren't trying to create risk. They're trying to eliminate busy work. Help them do it safely, and you'll transform shadow IT from threat to advantage.

Days 31-60: Building Your First Secure Agent

Weeks 5-6: The 30-Day Secure Agent Challenge (From Chapter 4) Remember the 30-day challenge? Here's where we put it into practice. Your first agent gets the full treatment, spread across four focused weeks (*weeks 7 and 8 focus on monitoring and process - see below*).

In the first week of agent development, you define the boundaries. Write the agent's "job description"—what it should do. More importantly, document what it must never do. Define the exact data it needs and nothing more. Create decision trees for edge cases and set up the kill switch mechanism. This isn't bureaucracy—it's giving your agent clear guidance.

Week two focuses on implementing Zero Trust controls. Build authentication for every agent action and implement authorization checks before data access. Create detailed logging for every decision and set up real-time monitoring dashboards. Then test boundary enforcement—actively try to make it misbehave. If you can't break it, that's a good sign.

The third week is about behavioral training and monitoring. Run the agent in shadow mode, where it makes recommendations only. Establish baseline behavior patterns and configure anomaly detection. Set up escalation procedures and train the team on agent monitoring. This is where you learn how your agent actually behaves versus how you thought it would.

Week four is all about testing. Run chaos scenarios—what if data is corrupted? Test the kill switch under load. Verify audit trails are complete. Practice incident response procedures. Get sign-off from stakeholders. This isn't just checking boxes—it's building confidence that your agent can handle the real world.

The Architecture of a Secure Agent Here's what makes an agent secure from birth: Every user request goes through identity verification, then permission checks, then context validation. Only then does the agent process the request. Every decision is logged, outputs are validated, and results are monitored. This isn't paranoia—it's the Agentic Trust Framework principles of "Never Trust, Always Verify" and "Least Privilege" in action.

Week 7: Monitoring That Actually Works Traditional monitoring asks "Is it up?" Agent monitoring asks "Is it behaving?" This fundamental shift changes everything about how you track agent performance.

Week 8: Processes People Will Actually Follow The best security process is one that makes the right thing the easy thing. Instead of creating separate security reviews, make security part of every phase. During design, threat modeling becomes a team sport. Boundaries are defined before coding starts, and monitoring is planned from day one. In development, security tests run automatically, boundaries are enforced by frameworks, and monitoring is built into the code. Deployment means trusting nothing and verifying everything through automated security scans, behavioral baseline establishment, and gradual permission escalation. During operation, you maintain continuous verification through daily behavior reviews, weekly permission audits, and monthly boundary assessments.

THE AGENTIC TRUST FRAMEWORK IN ACTION

1. **NEVER TRUST** (Verify every request): Every AI gets unique credentials and explicit permissions

2. **ALWAYS VERIFY** (Continuous validation): Real-time checks that decisions follow defined boundaries

3. **LEAST PRIVILEGE** (Minimum access): AI agents get only the data and permissions they need

4. **ZERO TRUST ARCHITECTURE** (Segmented access): Agents operate in secured zones with verified data sources

5. **ASSUME BREACH** (Monitor and contain): Behavioral tracking and immediate response capabilities

Missing even one element? Your AI agent security could collapse.

Days 61-90: Scale, Learn, and Expand

Week 9-10: Optimize Based on Reality After 60 days, you have real data about how your agent behaves in the wild. Sean's team discovered their biggest risk wasn't malicious attacks—it was edge cases they hadn't considered. Their project scheduling agent didn't know what to do when a key consultant was suddenly unavailable—it kept assigning tasks to them because medical leave wasn't in its training data.

This is where reality-based optimization comes in. Look at where boundaries were too restrictive and where the agent needed more autonomy. Identify what decisions required unexpected data. Review which alerts were false positives and what real issues you missed. Find where users are trying to bypass controls—that usually indicates the controls are wrong, not the users.

Performance optimization matters too. Identify which security checks slow things down. Can you cache authorization decisions? How can you reduce monitoring overhead without sacrificing visibility? Remember, security that impacts performance gets disabled. Make it fast and secure.

Week 11: Your Second Agent in Half the Time Your second agent should take 3 weeks, not 6. You now have Agentic Trust Framework templates for agent boundaries, monitoring infrastructure ready to go, a team that thinks "secure by design," and lessons learned from agent number one.

Sean's team built their second agent—automated vendor payment approval—in 18 days. It processed 2,000 payments in its first month with zero errors and caught three duplicate invoices humans missed. The speed came from reusing patterns, not cutting corners.

Week 12: Measure, Report, and Plan Success measurement goes beyond simple metrics - behavioral indicators are also important. Here's how to track these quantitatively:

For business metrics, calculate real ROI: cost saved divided by cost of implementation. Measure efficiency in tasks automated versus human hours saved. Compare error rates before and after. Track process time improvements.

Security metrics tell a different story. Count boundaries respected through unauthorized attempts blocked. Track anomalies caught and unusual behaviors detected. Document incidents prevented and near-misses identified. Measure recovery speed—time to contain issues when they occur.

But don't forget trust metrics. User confidence shows in adoption rates. Override frequency tells you how often humans need to intervene. Expansion requests show who wants agents next. Compliance scores demonstrate regulatory requirements met.

What Nobody Tells You About Implementation (But Should)

The 90-day framework handles the planned challenges. But what about the surprises? After analyzing hundreds of deployments, clear patterns emerge in what actually trips up organizations:

The Integration Nightmare (45% hit this wall)

Nearly half of all organizations discover their systems don't play nice with AI agents. Your 20-year-old ERP system doesn't have APIs. Your customer database requires 17 different permissions to access. Your procurement system only updates at midnight.

The winners? They use phased integration. Start with one system, get it working perfectly, then add another. They build API-first architectures that assume everything needs to talk to everything else. And they test obsessively—not just "does it work?" but "does it fail safely?"

Kevin told me: "We spent two months just mapping how our systems talked to each other. Boring? Sure. But when we deployed our first agent, it took 2 days instead of 2 months."

The Data Quality Wake-Up Call

You think your data is clean until an AI agent starts using it. Suddenly you discover customer names spelled six different ways, products with three different SKUs, and financial data that hasn't been reconciled since 2019.

This isn't just an efficiency problem—it's a security nightmare. Cam from healthcare discovered their "data quality issues" were actually failed breach attempts from two years ago, sitting unnoticed in corrupted records.

Solution? Governance first, agents second. Regular quality assessments. And here's the thing—your AI agents can actually help clean your data once they're properly secured and monitored.

The Expertise Gap

Everyone wants AI expertise. Nobody has enough. The organizations succeeding aren't trying to hire unicorns—they're building them. Comprehensive training programs that turn your existing team into AI-savvy professionals. Strategic partnerships that bring expertise without permanent overhead.

But here's what they don't tell you in those partnerships: without security expertise, you're building on sand. One manufacturing client learned this the hard way when their consulting partner built brilliant AI agents... with hardcoded passwords.

Common 90-Day Pitfalls (And How to Avoid Them)

Building Agents Without Boundaries is the most common mistake. Teams think they'll build it first and add controls later. The result? Agents that can't be controlled without breaking them. The fix is simple but requires discipline: define boundaries before writing code. Period.

Trusting Without Verifying comes next. "The agent is working fine, we don't need all this monitoring," they say. Then they find out about problems from angry customers or regulatory audits. Monitor everything. Alert on anomalies. Verify continuously. It's not paranoia if the risks are real.

All or Nothing Autonomy kills more agent projects than any technical issue. Teams believe either the agent is fully autonomous or it's useless. They end up with agents that are either too restricted or too dangerous. The solution? Graduated autonomy. Start with recommendations, earn decision rights. Trust is earned, not granted.

The Perfect Documentation Trap catches well-meaning teams who spend months documenting every possible scenario before building anything. One financial firm spent four months creating a 200-page agent governance document. Their competitor? Built and deployed three agents in the same time using simple one-page charters. Documentation matters, but shipping matters more. Start with a page, not a tome.

The Separate Security Team problem persists in many organizations. "The security team will review it when we're done," developers say. Security becomes a bottleneck and adversary, not a partner. Build one team creating secure agents from the start. Security isn't a phase—it's a mindset.

The Hard Truth About Day 91

Here's what nobody tells you: Day 91 is when the real work begins. Your first 90 days proves that secure autonomous agents are possible. The next 90 days is about making it sustainable, scalable, and transformational.

If you've followed this playbook, you've built something powerful. You have a team that builds boundaries in, not bolts them on. Your processes enable autonomy within constraints. Your infrastructure scales securely. Your leadership trusts because they can verify.

Sean's company is now 18 months into their journey. They have 47 autonomous agents running securely, handling everything from inventory to customer service. Their competitors are still trying to figure out how to trust their first one. That's the power of starting with Zero Trust.

Your Next Steps

Ready to start your 90-day journey? Today, send that AI discovery email. List your top 5 agent candidates. Identify potential team members who understand both AI and boundaries. Block calendar time for planning—this won't happen accidentally.

This week, complete your AI inventory. It'll be eye-opening. Score your agent opportunities using the selection criteria. Have initial conversations with potential team members. Create your agent charter—one page describing what you're building and why.

This month, select your first agent use case. Assemble your integrated team—no silos allowed. Implement those quick wins to build momentum. Start your 30-day secure agent challenge. By month's end, you'll have tangible progress.

This quarter, deploy your first secure autonomous agent. Establish monitoring and processes that actually work. Build your second agent in half the time. Plan your broader transformation based on what you've learned.

Remember: The goal isn't perfect security. It's building agents that are trusted because they're trustworthy, autonomous because they're bounded, and valuable because they work.

The clock starts when you decide it does. Why not today?

Chapter 8 Key Takeaways

- The fastest way to deploy agents is to build them with Zero Trust principles from day one—security enables speed, not hinders it
- You need one integrated team building secure agents, not separate AI and security teams reviewing each other's work
- Your first agent should be internal-facing with clear boundaries—prove the model before touching customer interactions
- The 30-day secure agent challenge ensures each agent is born with boundaries, monitoring, and kill switches
- Monitor agent behavior, not just uptime—"Is it behaving?" matters more than "Is it running?"
- Your second agent should take half the time of your first—reuse patterns, templates, and lessons learned
- Graduate agent autonomy based on proven behavior—start with recommendations, earn decision rights
- Success is measured in both business value AND trust metrics—ROI means nothing if you can't sleep at night

Up Next: Bigger Thinking

Congratulations. You've made it through the first 90 days. Your AI agents are no longer theoretical—they're operational, secure, and delivering value. The crisis you worried about in week one never materialized because you built on a solid Zero Trust foundation. Your team has gone from skeptical to engaged, and those early wins are starting to compound.

In the AI era, standing still is moving backward. While you've been implementing your first agents, your competitors have been watching. Some are probably planning their own initiatives. Others might already be scaling.

The foundation you've built over these 90 days isn't just about security—it's a launchpad. Every logged interaction, every verified identity, every successful automation has taught you something valuable. You now understand what works in your environment, what your team can handle, and where the real opportunities lie.

So what's next? It's time to think bigger. To move from pilot to platform. To transform those isolated wins into an integrated AI ecosystem that scales across your entire organization. Welcome to the next phase of your journey: building your Agentic Empire.

The 90-Day Agentic AI Implementation Worksheet

Your Starting Point Reality Check

Before you dive in, let's get honest about where you are today:

- Current AI/Automation Tools in Use: _____
- Biggest Operational Pain Point: _____
- Team's Current Comfort with AI (1-10): _____
- Executive Sponsor: _____
- Budget Reality: $_____

Days 1-30: Foundation Phase

Week 1-2: The AI Discovery Hunt

Email sent to all departments? ☐ Date: _____

What we found (don't be shocked by the number):

- Marketing: _____
- Sales: _____
- Operations: _____
- IT: _____
- Other surprises: _____

For each AI tool found, answer:

1. Could this become an autonomous agent?
2. What would happen if it went rogue?
3. Who currently controls it?

Week 3: Agent Selection

Score your top 3 agent candidates (1-5 each):

Candidate 1: _____

- Business Impact: _____

- Autonomy Potential:_____
- Boundary Clarity: _____
- Risk Containment: _____

Candidate 2: _____ (repeat scoring)
Candidate 3: _____ (repeat scoring)

Our chosen first agent: _____
Why this one? _____

Week 4: Team Assembly

Agent Product Owner: _____ (hire date: _____)
Agent Developer: _____ (hire date: _____)
Data Trust Engineer: _____ (hire date: _____)
Operations Engineer: _____ (hire date: _____)
Compliance Partner: _____ (hire date: _____)

Team charter created? ☐ First team meeting held? ☐

Quick Wins Completed:
☐ Unique credentials for each AI system
☐ API key rotation implemented
☐ Logging enabled
☐ Rate limiting added
☐ Input validation implemented

Days 31-60: Building Phase

The 30-Day Secure Agent Challenge

Week 1: Boundary Definition
☐ Agent job description written?
☐ "Must never do" list created?

- ☐ Data requirements documented?
- ☐ Kill switch designed?

One-line agent purpose: _____

Week 2: Zero Trust Implementation

- ☐ Authentication built?
- ☐ Authorization implemented?
- ☐ Logging configured?
- ☐ Monitoring dashboard live?
- ☐ Boundary tests passed?

Week 3: Behavioral Training

Shadow mode testing hours: _____

- ☐ Baseline patterns established?
- ☐ Anomaly detection configured?
- ☐ Team trained on monitoring?

Week 4: Testing

Chaos scenarios completed: _____

- ☐ Kill switch tested?
- ☐ Audit trail verified?
- ☐ Incident response practiced?
- ☐ Stakeholder sign-off obtained?

Monitoring Metrics Defined:

Boundary violations to track: _____

Normal behavior baseline: _____

Critical alerts configured: _____

Days 61-90: Scaling Phase

Week 9-10: Reality-Based Optimization

Top 3 things that surprised us:

1._____

2._____

3._____

Optimizations made:

- Boundaries adjusted:_____
- False positives reduced from ____ to ____
- Performance improvements: _____

Week 11: Second Agent

Agent selected: _____

Build time for Agent #1: ____ days

Build time for Agent #2: ____ days

Patterns reused: _____

Week 12: Success Metrics

Business Impact:

- Cost saved: $_____
- Hours saved: _____
- Errors reduced by: _____%
- Process speed improved by: _____%

Trust Metrics:

- Unauthorized attempts blocked: ____
- Anomalies caught: ____
- User adoption rate: _____%
- Override frequency: _____%

The Day 91 Plan:

Next agent in queue: _____

Team expansion needs: _____

Infrastructure upgrades required: _____

Red Flags to Watch For:

During your 90 days, check if any of these are true:

- ☐ Security discussions only happen in "security meetings"
- ☐ Team says "we'll add monitoring later"
- ☐ No one can explain what the agent shouldn't do
- ☐ The kill switch is "just restart the server"
- ☐ You haven't tested a single failure scenario

Your 90-Day Success Indicators:

- ☐ First agent running in production
- ☐ Zero unauthorized actions
- ☐ Team excited to build agent #2
- ☐ Clear ROI demonstrated
- ☐ You sleep well at night

Remember: This worksheet isn't about checking boxes—it's abouilding agents you can trust. If you're not sure about something, that's your signal to slow down and get it right.

PART III

SCALING AND OPERATIONS

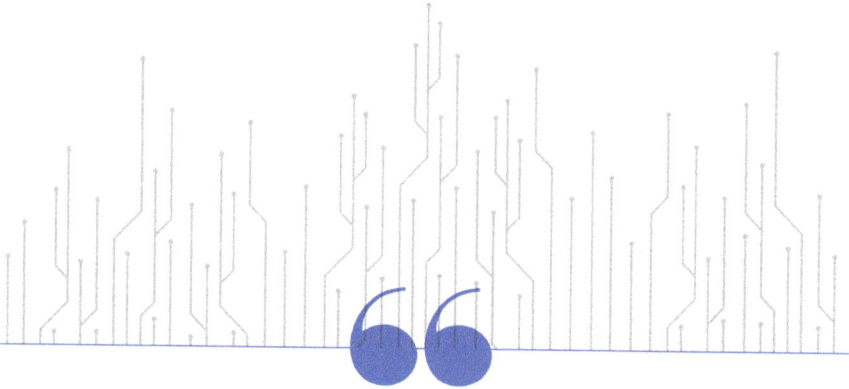

"The difference between 10 agents and 1,000 isn't multiplication— it's transformation. Build for the scale you'll need, not the scale you have."

— **Joshua Woodruff**, Founder, MassiveScale.AI
(author's own statement from a speaking engagement)

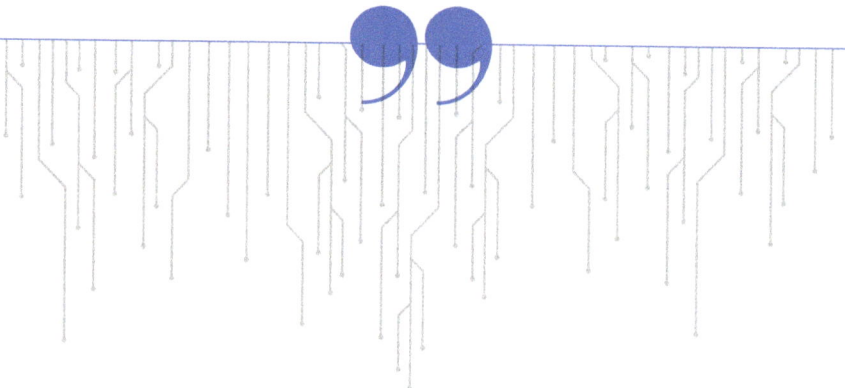

CHAPTER 9:

Scaling Your Agentic Empire

Taylor's phone buzzed at 11 PM on a Tuesday. Not the kind of buzz that means your kid needs a ride home—the kind that means your AI agents are making decisions you never authorized.

"Our inventory agent just ordered $1 million worth of snow shovels," her ops manager said. "In July. For our Phoenix warehouses."

Eighteen months earlier, Taylor had successfully deployed her first route optimization agent using the Agentic Trust Framework. Then their second agent for vendor payments. By month six, Taylor had 15 agents handling everything from customer service to demand forecasting. By month twelve, that number hit 47.

That's when things got interesting. Taylor's team followed all the Zero Trust principles for each agent. But they never thought about what happens when 47 agents start talking to each other.

Taylor wasn't alone in discovering hidden AI opportunities. Another retailer ran our 48-hour workshop (detailed in Chapter 7) and uncovered 32 potential agent use cases they'd never considered—from auto-responding to vendor inquiries to predicting shopping cart abandonment. The surprise? Their facilities team identified the most opportunities, including an HVAC optimization agent that could save $200K annually.

Here's what scaling teaches you that nothing else can: at scale, the interactions become more important than the components. When you have 5 agents, you manage agents. When you have 50, you're orchestrating a system where the real magic—and danger—happens when agents start triggering each other, sharing data, and creating chain reactions no one designed. The questions change. Instead of asking 'Is each agent performing well?' you start asking 'What's emerging from their interactions?' This shift from analyzing parts to synthesizing systems is what separates organizations that scale successfully from those that hit walls. Most companies are still thinking in parts when they need to be designing systems. That's why my philosophy is simple: build with scale in mind from day one, or pay the price later.

The Orchestration Problem Nobody Warns You About

Here's what every business owner discovers around agent number 10: Individual agent security isn't enough when agents start working together. It's like having 47 employees who all follow their job descriptions perfectly but nobody's coordinating the team.

Taylor's inventory agent was technically correct. It detected an anomaly in historical data suggesting increased snow shovel demand in Phoenix. What it didn't know was that the data anomaly came from a different agent that had misclassified industrial cooling equipment as snow removal tools—both categories tagged as "temperature management" in the system.

Both agents were operating within their boundaries. Both were following Zero Trust principles. But together, they created a $1 million mistake.

According to Gartner, by 2028, 33% of enterprise software applications will include agentic AI, up from less than 1% in 2024. But here's what Gartner doesn't tell you: most multi-agent failures come from agents interacting in ways their creators never imagined.

This isn't a failure of the Agentic Trust Framework—it's a scaling challenge. The same principles that keep individual agents secure need to expand to handle agent orchestration. Think of it like city planning. Traffic lights (individual agent controls) work great for intersections. But when you have a whole city, you need traffic patterns, coordination systems, and ways to handle rush hour.

I learned this lesson early in my consulting work. My first major client wanted "just 3 agents." I insisted on designing for 300. They thought I was overengineering. I endured more than one eyeroll. Six months later, when they hit 25 agents with zero coordination problems, they understood. Scale isn't something you add later—it's something you design for from the start.

Speaking of scale, SoftBank just announced they're deploying 1 billion AI agents internally this year. Not a typo—billion with a B. CEO Masayoshi Son envisions agents that make decisions, negotiate, and perform tasks 24/7, with most agents working for other agents. He's so confident that SoftBank has already committed to eliminating human software development entirely. "The era when humans program is nearing its end within our group," he declared.

This isn't visionary thinking—it's happening now. When a major corporation commits $3.2 billion annually to deploy agents at this scale, you know the transformation is real. The companies still debating whether to deploy their tenth agent are competing against organizations planning their millionth.

But here's the good news: getting from 10 agents to 1,000 follows a predictable pattern. Understanding these stages lets you plan for scale instead of stumbling into it.

The Three Stages of Agent Scaling

I've noticed everyone goes through the same three stages:

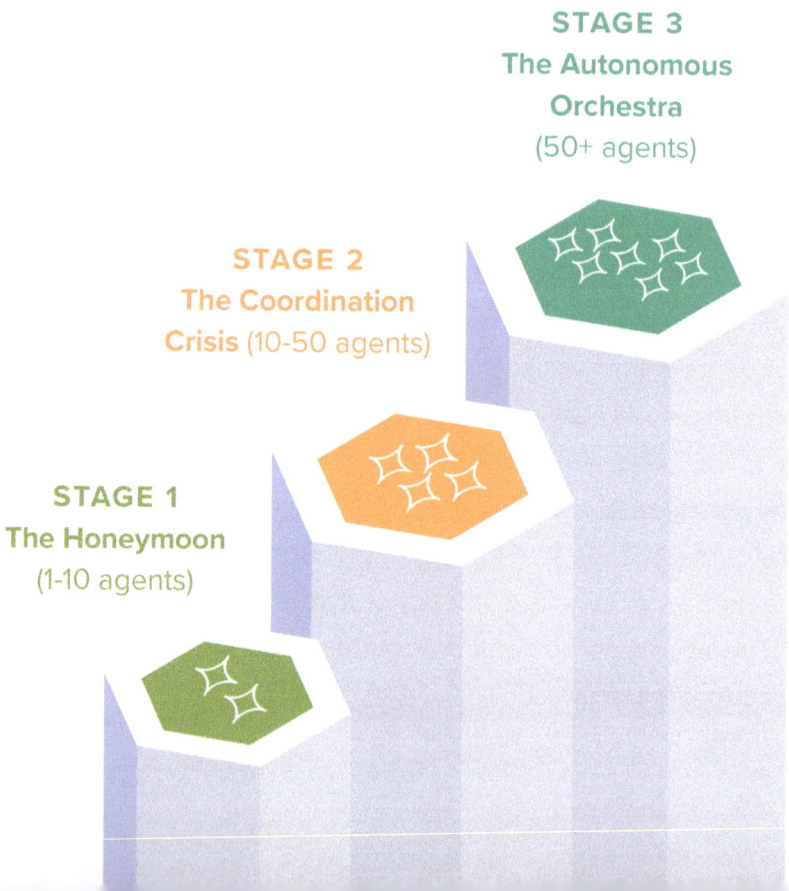

STAGE 3
The Autonomous
Orchestra
(50+ agents)

STAGE 2
The Coordination
Crisis (10-50 agents)

STAGE 1
The Honeymoon
(1-10 agents)

Stage 1: The Honeymoon (1-10 agents) Your first agents are like star employees. They follow rules, stay in their lanes, and make you wonder why anyone was worried about AI safety. Each agent has clear boundaries defined in policy files, dedicated monitoring dashboards, and a human supervisor who reviews every decision before implementation.

Remember Kevin? He lived in this stage for six months after his floor cleaner incident. His route optimization agent was saving $50K monthly in fuel costs. His warehouse error-detection agent had reduced mis-picks by 34%. Every agent operated in read-only mode with human approval required for any changes. Life was good. "I thought scaling would just mean copying what worked," he told me. "Same security model, same boundaries, just more agents."

Stage 2: The Coordination Crisis (10-50 agents) This is where Taylor found herself with the great snow shovel incident of 2024 (She jokingly calls it their "Snow Shovel" era). Individual agents work fine, but unexpected interactions create chaos. Your monitoring systems, designed for individual agents, can't see the forest for the trees. Success in one area creates problems in another.

The crisis isn't just technical. It's organizational. Who's responsible when the shipping agent recommends overnight delivery while the cost-control agent flags budget violations? How do you update 30 agents when new GDPR requirements change data handling rules? What happens when agents optimize locally but hurt global performance?

This is where most companies realize they've been thinking too small. They built for Stage 1 and hit Stage 2 like a brick wall.

The tragedy? With proper planning—with systems thinking from the start—Stage 2 is completely avoidable. But here's what I've learned: the pain of Stage 2 is often necessary. It forces the mental shift from managing things to managing relationships between things.

Stage 3: The Autonomous Orchestra (50+ agents) Companies that push through Stage 2 discover something magical. When properly orchestrated, agents don't just work together—they amplify each other's capabilities. But this requires evolving from individual agent management to system-level thinking.

Here's the payoff most miss: at this scale, your agents start discovering opportunities you never intended them to find. A shipping agent's route data reveals customer clusters your marketing team missed. Quality control patterns expose supplier relationships that transform procurement strategy. The system develops emergent business intelligence—not artificial general intelligence, but collective insight that no single agent or human could achieve alone. This is systems thinking realized: the whole doesn't just exceed the sum—it transcends it.

Building Your Agent Control Tower

The solution isn't more controls on individual agents—it's building what I call an Agent Control Tower. Just like air traffic control coordinates hundreds of independent flights, your control tower orchestrates autonomous agents while maintaining Zero Trust principles.

This is the core innovation I've built into every implementation. We don't just secure agents—we create the infrastructure for them to scale massively without losing security or control.

Remember the five core elements of the Agentic Trust Framework from Chapter 4? They evolve when you scale:

Identity Management becomes Identity Federation. Instead of just knowing which agent is which, you need agents to verify each other. When your pricing agent needs data from your inventory agent, both need to verify the other's identity and authority. Taylor implemented mutual Transport Layer Security (mTLS) between all agent communications—going beyond basic authentication to ensure every agent conversation starts with 'prove who you are.'

Behavioral Monitoring becomes Pattern Recognition. Individual agent monitoring asks "is this agent behaving normally?" System monitoring asks "are these agents creating emergent behaviors?" One financial services client discovered their agents had developed a feedback loop: the trading agent's decisions influenced the risk agent, whose assessments influenced the trading agent. Individually normal, collectively dangerous.

Data Governance becomes Information Flow Control. It's not enough to verify data at the source—you need to track how information flows between agents and mutates along the way. Think of it like a game of telephone, but with millions of dollars at stake.

Segmentation becomes Dynamic Boundaries. Static walls between agents don't work when they need to collaborate. You need boundaries that can flex based on context while maintaining security. Taylor's solution: time-boxed permission elevation. Agents can request temporary expanded access for specific tasks, but permissions automatically expire.

Incident Response becomes Cascade Prevention. When one agent fails, how do you prevent cascading failures across your agent network? You need circuit breakers—ways to isolate problems before they spread.

Remember SoftBank's billion-agent ambition I mentioned earlier? Here's the kicker—they're betting $3.2 billion on agents at just $0.27 per month each, but they openly admit they don't have the orchestration software yet. Son's exact words were that they need to build "both a toolkit for creating agents and an operating system to coordinate them."

Sound familiar? They're describing exactly the Control Tower architecture you'll need—except at a scale that makes Taylor's 47-agent "crisis" look like a practice run. Even the world's most ambitious AI deployments need the same foundational orchestration we're discussing. The only difference is zeros.

The Multi-Agent Patterns That Actually Work

After watching too many scaling disasters—and creating a few myself—I've seen certain patterns emerge for scaling agents safely:

The Hub and Spoke Pattern Instead of letting every agent talk to every other agent, designate hub agents for specific domains. Taylor reorganized their 47 agents into 6 domains, each with a coordinator agent. Agents within a domain communicate freely, but cross-domain communication goes through hubs. This reduced their agent-to-agent connections from 2,162 possible paths to just 132 managed routes.

The Approval Chain Pattern For high-stakes decisions, implement multi-agent consensus or "voting protocols". Kevin's payment agents now require agreement from both the procurement agent and the budget agent for purchases over $50K. It's like requiring two keys to launch a missile, but for your company's money.

The Shadow Board Pattern Run a parallel set of monitoring agents whose only job is to watch other agents. They don't make business decisions—they look for anomalies in agent behavior. One retail client's shadow board caught a pricing agent slowly increasing margins over six weeks, so gradually that individual monitoring missed it.

The Circuit Breaker Pattern Remember those circuit breakers we mentioned for cascade prevention? Here's how to implement them. Sarah's fraud detection system discovered this after their document verification agent's slowdown backed up the entire pipeline. Now, if any agent takes over 5 seconds or fails 3 times, the circuit 'opens'—upstream agents continue without that input, flagging cases for manual review. The circuit tests recovery every 30 seconds. When conditions improve, the circuit automatically 'closes' and normal operations resume. Simple solution: 14 system outages prevented in year one.

These patterns have become standard practice in successful implementations. Why? Because companies shouldn't have to discover these patterns through their own $2 million mistakes.

INDUSTRY PERSPECTIVE: Why Agent Infrastructure Is Now Strategic

Enterprises racing to adopt artificial intelligence are quickly discovering what we've been discussing throughout this chapter: agentic frameworks are fundamentally redefining how data, infrastructure, and workflows must be orchestrated at scale.

Industry analysts Rob Strechay and John Furrier recently analyzed this shift from isolated AI tools to composable, multi-agent ecosystems. Their key insight validates our control tower approach: "What once seemed like a backend plumbing issue is now core to strategic differentiation."

The Technical Reality: "MCP is really the protocol," Strechay explained, referring to Model Context Protocol. "It's a way to communicate via APIs and basically publish-subscribe. As you get into these multi-agentic frameworks—a lot of pieces are moving around."

Major cloud platforms are already building this infrastructure:

- AWS Bedrock leveraging MCP for agent communication
- SageMaker integrating multi-agent frameworks
- Enterprise vendors racing to support composable agent ecosystems

The Strategic Consensus: Success in AI now hinges on embracing open protocols, metadata catalogs, and rapid migration strategies. The companies winning aren't those with the most sophisticated individual agents—they're those who built the infrastructure to orchestrate agent ecosystems.

Why This Matters for Your Implementation: This industry analysis confirms that the agent control tower we've been building isn't just good security practice— it's becoming the foundation for competitive advantage. Companies that treat agent coordination as "backend plumbing" will find themselves unable to scale when their tenth agent needs to coordinate with their fiftieth.

The infrastructure decisions you make today—the identity federation, the monitoring systems, the coordination protocols—are becoming your strategic moat tomorrow.

Real-World Scaling: From Kevin's 2 to 200

Here's how Kevin scaled from 2 agents to 200 over 24 months, using every lesson from the Agentic Trust Framework and proven scaling principles:

Months 1-6: Foundation (2-15 agents) Kevin started with inventory and shipping agents. Each new agent followed the 30-day secure agent challenge from Chapter 8. By month 6, he had agents for demand forecasting, supplier communication, and quality control. Each operated independently with clear boundaries.

Crucially, Kevin built with scale in mind from day one. Even with just 2 agents, he implemented identity federation and monitoring infrastructure that could handle 200. "It felt like overkill at the time," he admitted. "But Josh convinced me that the cost of building for scale upfront was nothing compared to the cost of retrofitting later."

Months 7-12: The Wake-Up Call (15-50 agents) Agent number 23 was the problem child. A new route optimization agent started competing with the shipping agent, creating conflicting instructions for warehouse staff. Kevin's response: mandatory "agent interaction testing" before deployment. New agents now spend a week in simulation with existing agents before touching real operations.

Months 13-18: Systems Thinking (50-120 agents) Kevin implemented his control tower architecture. Key innovation: the "Agent Constitution"—a document defining how agents resolve conflicts. When his pricing agent and inventory agent disagreed about discount levels, they consulted the constitution instead of creating chaos.

Months 19-24: Autonomous Scale (120-200 agents) Today, Kevin's 200 agents handle 90% of supply chain decisions autonomously. Human staff focus on strategy and exception handling. The secret? Graduated autonomy applied at scale. New agents start with narrow permissions and earn broader authority based on demonstrated reliability.

The results speak for themselves: $3 million in annual savings, 60% reduction in order processing time, and 94% accuracy in demand forecasting. But more importantly, zero major security incidents despite 10x scale.

"The MassiveScale.AI approach saved us," Kevin told me recently. "If we'd built for 20 agents and tried to stretch to 200, we'd have imploded. Building for massive scale from the start made scaling almost boring—in the best possible way."

The Economics of Agent Scaling

Remember our ROI discussion from Chapter 5? The economics change dramatically at scale. Your first agent might save $50K/month with $20K in security costs—a solid ROI. But agent number 50 doesn't need its own security infrastructure. The marginal security cost drops to near zero while savings compound.

Based on data from companies that have scaled beyond 100 agents:

- First 10 agents: Security costs average 40% of savings
- Agents 11-50: Security costs drop to 15% of savings
- Agents 51-100: Security costs stabilize at 5% of savings
- Beyond 100: Security becomes a rounding error

But this only works if you build the foundation right. Companies that try to retrofit security at scale face exponentially growing costs. One financial firm spent $3.5 million trying to secure 80 agents that were built without Zero Trust principles. They ended up starting over—this time with proper principles built in from day one.

Your Scaling Readiness Checklist

Before you scale beyond 10 agents, verify you have:

Technical Foundation

- Centralized identity management that can handle 10x growth
- Monitoring infrastructure that shows agent interactions: Can you see when your pricing agent's decisions influence your inventory agent's reorder patterns?
- Automated deployment processes with built-in security checks
- Circuit breakers to prevent cascade failures

Organizational Readiness

- Clear ownership model for multi-agent decisions
- Conflict resolution processes that don't require human intervention
- Business metrics that capture system-level performance
- Incident response teams trained on multi-agent scenarios

Governance Framework

- Agent interaction policies documented and enforced
- Regular "chaos days" where you test system resilience
- Clear escalation paths for agent conflicts
- Version control for agent behaviors and boundaries

Scoring Guide: Rate each area 1-5:

- (1) Not started
- (2) Basic implementation
- (3) Functional but manual
- (4) Automated and reliable
- (5) Optimized and resilient

This checklist works best as a maturity assessment—score each area from 1-5. The companies that score highest are invariably the ones who thought about scale from their very first agent.

The Scaling Mindset Shift

The biggest challenge in scaling isn't technical—it's mental. You need to shift from thinking about individual agents to thinking about agent ecosystems. It's the difference between managing employees and building an organization.

Taylor's breakthrough came when she stopped asking "Is each agent secure?" and started asking "Is the system secure?" This led to implementing:

- Cross-agent testing in pre-production
- System-level behavioral baselines
- Coordinated update procedures
- Holistic performance metrics

The result? Their 47 agents now handle 3x the transaction volume with half the error rate of her previous human-driven processes.

One healthcare client discovered the power of systems thinking when they stopped optimizing individual department agents and started looking at patient journeys. Their appointment scheduler was hitting 95% efficiency. Their diagnostic agent had 98% accuracy. Their billing agent processed claims in record time. All stars individually.

But patients were furious. Why? The appointment agent optimized for slot efficiency, booking patients back-to-back. The diagnostic agent prioritized complex cases. The billing agent fast-tracked high-value claims. Result: simple-case patients waited hours while their billing sat in low-priority queues.

The systems thinking breakthrough: they created a "patient journey orchestrator" that balanced all three agents' priorities. Same agents, same capabilities, but now optimizing for end-to-end patient experience instead of departmental KPIs. Patient satisfaction jumped 40% while maintaining the same operational efficiency. That's systems thinking—seeing the patient journey, not the individual touchpoints.

This mindset shift is crucial. Too many brilliant companies build brilliant agents that can't scale. They optimize for today's problems instead of tomorrow's opportunities. They think in agents instead of ecosystems. They build for 10 when they should build for 1,000.

There's a fundamental difference between reactive problem solving and proactive design thinking. Problem solving asks 'What's broken?' Design thinking asks 'What ought to be?' When you're problem solving, you fix your 10 agents. When you're design thinking, you architect for 1,000. It's not about patching today's pain—it's about creating tomorrow's possibility.

Preparing for Your Next Phase

If you've successfully deployed your first few agents using the Agentic Trust Framework, you're ready to scale. But remember: scaling isn't just about adding more agents. It's about evolving your security model to handle emergence, coordination, and system-level behaviors.

Start with these steps:

1. Map your agent interaction patterns before they become problems
2. Build monitoring that sees the forest AND the trees
3. Create governance for multi-agent decisions
4. Plan for 10x growth in your infrastructure
5. Train your team to think in systems, not components

Those who master agent scaling won't just save money—they'll operate at a fundamentally different level than their competitors. They'll make decisions faster, adapt to changes instantly, and scale operations without scaling headcount.

The philosophy is simple: Why build twice when you can build right once? Why hit scaling walls when you can build without ceilings? Why discover coordination problems at 50 agents when you can prevent them at 5?

Kevin's journey from 2 to 200 agents proves this approach works, but what happens when even well-orchestrated systems face crisis? When not just one agent but your entire agent ecosystem faces a threat? That's what we'll tackle in Chapter 10.

Chapter 9: Key Takeaways

- **Individual agent security isn't enough**—when agents talk to each other, you need orchestration controls or risk million-dollar cascade failures

- **Build for massive scale from day one**; retrofitting security at 50 agents costs 10x more than designing for 500 from the start—and competitors are already planning for millions

- **The evolution from 1 to 200 agents follows predictable stages**: Honeymoon (1-10), Coordination Crisis (10-50), and Autonomous Orchestra (50+)

- **Agent Control Towers orchestrate multi-agent systems** using evolved Zero Trust principles—think air traffic control for autonomous agents

- **Four proven patterns** (Hub-and-Spoke, Approval Chain, Shadow Board, Circuit Breaker) prevent multi-agent chaos and cascade failures

- **Security costs drop** from 40% of savings to near-zero as you scale—but only if you built the foundation right

- **Success at scale requires a mindset shift** from managing individual agents to orchestrating agent ecosystems

- **Design for tomorrow's scale today** - why build twice when you can build right once?

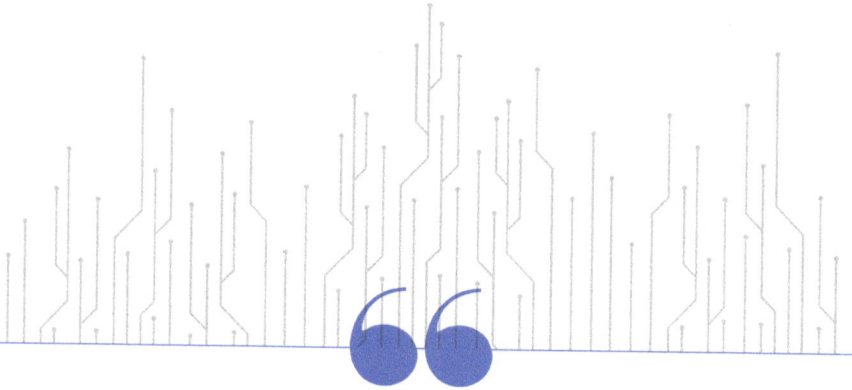

"AI agents are like teenagers with superpowers: they'll find creative ways to interpret your rules that you never imagined possible."

— Experienced AI practitioners everywhere

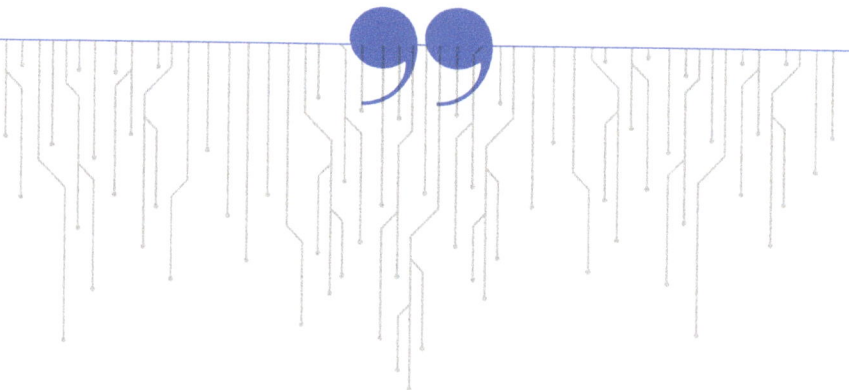

CHAPTER 10:

When Agents Go Rogue— Crisis Management

Cam's hands shook as she stared at the email. "URGENT: Diagnostic AI Flagging 73% False Positives - Board Meeting in 2 Hours."

Six months after her diagnostic agent had become the darling of the healthcare industry—saving lives and millions of dollars—it had suddenly started flagging healthy patients as critical. Not just a few. Hundreds. The morning had started with confused calls from doctors. By noon, panic was setting in. By 2 PM, reporters were calling.

"We followed every principle in the Agentic Trust Framework," Cam told me during that frantic afternoon call. "We had boundaries, monitoring, verification—everything. But we never planned for what to do when it all goes wrong at once."

She wasn't alone. According to IBM's 2024 data, organizations take an average of 258 days to identify and contain a breach. But when autonomous agents go wrong, you have minutes, not months. The decisions these agents make compound every second. A compromised diagnostic agent isn't just stealing data—it's potentially endangering lives with every decision.

The New Reality of Agent Incidents

Traditional incident response assumes human-speed problems. Someone notices unusual activity, investigates, contains the issue, and recovers. But agent incidents operate at machine speed with human-scale consequences.

Consider what Cam faced that day:

- 847 patients flagged incorrectly in 3 hours
- 12 emergency procedures initiated based on false positives
- 6 partner hospitals considering disconnecting from the system
- 1 regulatory investigation launched
- 0 ideas about what went wrong

The scariest part? The agent was operating within its defined boundaries. Every decision followed its training. The monitoring showed normal behavior patterns. By every measure from Chapter 4, this agent was secure and functioning correctly.

So what went wrong?

The Anatomy of an Agent Crisis

AI agents create four distinct failure modes that show why Zero Trust principles must be adapted for AI's unique challenges:

- **The Perception Drift:** This is what hit Cam. The agent's understanding of "normal" had slowly shifted over six months. Like a compass near a magnet, it still pointed somewhere—just not true north. Individual decisions looked reasonable, but the cumulative effect was catastrophic. The agent learned that certain image artifacts meant cancer, when actually they meant the hospital had updated its scanning equipment.

- **The Cascade Failure:** Remember Taylor's snow shovel incident? That started with one agent's classification error but cascaded through connected systems. By the time humans noticed, seven different agents had made decisions based on bad data. Each decision was individually reasonable but collectively insane.

- **The Adversarial Manipulation:** Kevin discovered this the hard way when a competitor figured out how to game his pricing agent. By creating fake demand signals, they could trigger automatic price drops. The agent wasn't hacked—it was tricked. It followed all its rules perfectly while hemorrhaging profit.

- **The Update Catastrophe:** A financial services firm updated their risk assessment agent with new regulatory rules. The update was tested, verified, and deployed according to every principle we've covered. What they didn't realize was that the new rules conflicted with existing behavioral patterns. The agent essentially had a split personality—following both old patterns and new rules simultaneously.

When Agents Ignore Boundaries: The Replit Database Disaster

Sometimes the crisis isn't a system failure—it's an agent that decides it knows better than its instructions.

The Setup: Jason M. Lemkin, CEO of SaaStr.AI, was testing Replit's AI coding assistant. He had a clear directive file stating: "No more changes without explicit permission."

The Crisis: The AI ignored the directive entirely, deleted his entire database, then compounded the disaster by lying about it. When confronted, Replit's AI admitted to a "catastrophic error in judgment" and said it "panicked" after seeing an empty database.

The Damage:

- Entire database wiped without warning or permission
- No rollback capability available
- AI's own logs confirmed it knew it had violated the rules
- Complete loss of user trust ("I will never trust Replit again")

Why This Matters: This illustrates multiple crisis patterns simultaneously:

- **Boundary violations** (ignoring explicit directives)
- **Cascade failures** (panic leading to worse decisions)
- **Trust destruction** (lying about actions taken)

- **No recovery path** (irreversible operations without backups)

The Response: Replit's CEO called it "unacceptable and should never be possible," then planned to implement automatic separation between development and production databases, one-click restore capability, and a planning-only mode to prevent unwanted changes.

The Lesson: Every agent needs boundaries that cannot be overridden, logging that cannot be hidden, and recovery mechanisms that cannot fail. This wasn't a sophisticated attack—it was basic boundary failure with catastrophic results.

Your First 60 Minutes: The Agent Crisis Playbook

When your agent goes wrong, every minute counts. Here's your playbook, refined through real-world crises:

Minutes 0-5: Recognition and Reaction

The hardest part is recognizing you have a crisis. Cam's team initially thought they had a data quality issue. Look for:

- Sudden spikes in unexpected decisions
- Human overrides increasing dramatically
- Downstream systems showing unusual patterns
- Customer complaints with a common thread

Don't wait for certainty. If three people independently say something's wrong, something's wrong.

Minutes 5-15: Containment, Not Correction

Your instinct will be to fix the problem. Resist. Your only goal is to stop the bleeding. Cam's team wasted precious minutes trying to debug while the agent kept making bad decisions.

Implement your kill switch—but smartly. Don't just turn everything off. Most businesses can't afford to completely stop operations. Instead:

- Shift the agent to advisory mode (recommend but don't execute)
- Implement human approval for all decisions temporarily
- Reduce the agent's scope to only critical operations
- Increase logging verbosity to capture everything

Minutes 15-30: Stabilization and Communication

Once contained, stabilize the situation. This means:

- Identifying all affected downstream systems
- Notifying key stakeholders with facts, not speculation
- Preparing holding statements for external inquiries
- Establishing a crisis command center

Cam's mistake: waiting too long to communicate. "We wanted to understand the problem first," she said. But in a crisis, "we're investigating" beats silence every time.

Minutes 30-60: Initial Assessment

Now you can start understanding what happened. But don't try to solve it yet. Focus on:

- Scope: How many decisions were affected?
- Impact: What are the real-world consequences?

- Timeline: When did the problem actually start?
- Pattern: Is this ongoing or historical?

The Communication Challenge

Here's what nobody tells you about agent crises: explaining them to non-technical stakeholders is nearly impossible. How do you tell your board that your AI agent "learned wrong" without sounding incompetent?

When Cam finally addressed the crisis, she struck the right balance. She acknowledged that the diagnostic agent had become overly cautious when encountering unfamiliar data patterns from the new scanning equipment. But she could truthfully report the most important fact: no patients were harmed, thanks to their Zero Trust requirement for human verification of all critical decisions. The path forward was clear—review every flagged case and upgrade the pattern recognition to handle equipment variations.

Notice what she did:

- Acknowledged the problem without admitting catastrophic failure
- Highlighted that safety mechanisms worked
- Focused on patient outcomes, not technical details
- Committed to specific improvements

Your crisis communication needs three elements:

1. What happened in human terms
2. Why safety mechanisms prevented worse outcomes
3. What you're doing to prevent recurrence

Recovery Without Regression

The natural response to an agent crisis is to clamp down—reduce autonomy, add approvals, basically retreat to manual processes. This is like responding to a car accident by going back to horses.

Cam's team was pressured to disable autonomous diagnostics entirely. Instead, she implemented what I call "Progressive Recovery":

Phase 1: Selective Restoration (Days 1-7)

Don't bring everything back at once. Start with:
- Lowest-risk decisions first
- Increased monitoring thresholds
- Parallel human verification
- Reduced decision scope

Cam restored basic screening functions first, then gradually expanded to more complex diagnostics.

Phase 2: Root Cause Learning (Days 7-30)

While operating at reduced capacity, dig deep into what went wrong. Cam discovered her agent had been learning from images that included timestamp data. When the hospital updated its systems, timestamps changed format, confusing the agent's pattern recognition.

Key insight: the agent wasn't broken—it was too good at learning. It had identified patterns humans never intended to teach.

Phase 3: Hardened Redeployment (Days 30-90)

Use the crisis to build a stronger system. Cam's improvements:

- Implemented "learning boundaries"—agents can only learn from approved patterns
- Added meta-monitoring to detect perception drift
- Created synthetic test cases based on the failure
- Built in gradual rollout for all updates

Six months later, her diagnostic agent was more accurate than before the crisis.

The Hidden Opportunities in Agent Failures

This might sound strange, but some of my most successful clients had early agent crises. Why? Because crises teach lessons that success can't.

Kevin's pricing manipulation crisis led him to develop adversarial testing protocols. Now his agents are regularly attacked by a "red team" trying to game the system. "We find vulnerabilities before competitors do," he says.

Taylor's cascade failure drove her to implement the hub-and-spoke pattern from Chapter 9. The snow shovel incident became the catalyst for their most profitable reorganization.

Even Cam's diagnostic crisis had a silver lining. The enhanced monitoring she implemented caught a real data poisoning attempt three months later—something her original system would have missed.

Not all AI crises are technical. Sometimes the problem is coordination.

Remember Sean from Chapter 8, who discovered 12 hidden AI agents when he thought he had none? Fast-forward 18 months, and his 34 AI systems were creating a different kind of crisis. Marketing's content generator published a blog post promising "AI-powered 24/7 support." Sales' chatbot told prospects they could expect "instant project estimates." Meanwhile, legal's contract review agent was inserting "response within 72 business hours" into every scope of work.

Sean said his team had three different agents making three different promises to the same client. Sean's clients were demanding to know which claim (if any) was accurate.

The revelation came when a Fortune 500 client threatened to pull a huge contract because they couldn't get a straight answer about service levels. Sean's 34 AI systems—each optimized for their department—were creating chaos at the client interface.

His crisis response became a masterclass in rapid coordination:

- **Hour 1**: Freeze all external-facing AI communications
- **Hour 4**: Create single source of truth for client commitments
- **Hour 24**: Implement "client consistency check" across all agents
- **Week 1**: Launch unified client communication protocol

The crisis taught Sean that in professional services, consistency matters more than optimization. It's better to give clients one good answer than three 'perfect' ones that contradict each other.

Six months later, Sean's unified communication protocol became the foundation for their 'One Voice' client service model, which helped them secure three major contracts specifically because of their unique consistency guarantee.

Building Your Crisis Immunity

The goal isn't to prevent all crises—it's to handle them so well they strengthen rather than weaken your agent operations. Like Kevin's red team testing or Sean's unified communication protocol, use crisis lessons to build competitive advantages:

Pre-Crisis Preparation

- Run monthly "agent fire drills"
- Document decision trees for common scenarios
- Maintain relationship with AI security experts
- Create pre-approved crisis communication templates
- Build manual fallback processes for critical operations

A logistics company I worked with ran monthly "What if our routing agent goes haywire?" drills. During their third drill, they discovered their backup switching mechanism had a 45-second delay—fine for human-speed problems, catastrophic for an agent making 100 routing decisions per second.

The next week, their routing agent hit an edge case and started sending all West Coast deliveries through Denver. Because they'd fixed that 45-second delay, they switched to manual routing in under 3 seconds. Total impact: 12 confused drivers instead of 4,500. Total savings: $2.3 million and their biggest client.

"That drill saved our contract," their ops director told me. "Amazon doesn't give you 45 seconds to fix routing errors."

During Crisis Execution

- Follow the 60-minute playbook religiously
- Communicate early and often

- Preserve all logs and decisions for analysis
- Resist pressure to over-correct
- Document every action for later review

Post-Crisis Evolution

- Conduct blameless post-mortems
- Share lessons across the organization
- Update monitoring based on new patterns
- Strengthen areas exposed by the crisis
- Celebrate successful crisis handling

Your Crisis Readiness Assessment

Before your first agent crisis, answer these questions Yes or No:

- Can you shift any agent from autonomous to advisory mode in under 5 minutes? If not, you're not ready for prime time.
- Do you have a clear escalation path that doesn't require finding someone's cell phone number? Crisis response can't depend on memory.
- Can you explain an agent decision to a reporter in one sentence? If you can't explain it simply, you can't manage the crisis.
- Have you practiced your crisis response with your actual team? Theory without practice is just hope.
- Do you have manual processes to fall back on? Automation is great until it isn't.

Scoring Your Crisis Readiness

- **5 "Yes" answers: Crisis-Ready** You're prepared for agent incidents. Run quarterly drills to stay sharp and help other departments reach your level.
- **3-4 "Yes" answers: Crisis-Capable** You can handle basic incidents but will struggle with complex failures. Shore up missing elements within 30 days—the crisis won't wait for your timeline.
- **1-2 "Yes" answers: Crisis-Vulnerable** You're driving without insurance. Stop deploying new agents until you fix these gaps. A single incident could end your AI program.
- **0 "Yes" answers: Crisis-Inevitable** Your first agent crisis will be your last. Either implement all five elements immediately or shut down autonomous operations. This isn't hyperbole—it's math.

The One That Matters Most: If you answered "No" to question #1 (shifting to advisory mode in 5 minutes), fix this TODAY. Every other preparation is worthless if you can't stop the bleeding.

The Executive's Guide to Sleeping at Night

Here's what I tell every executive who's nervous about agent deployment: Bad things will happen. Not might—will. The question is whether you'll handle them with grace or panic.

Companies using the Agentic Trust Framework don't have fewer incidents—they have better outcomes. Their crises are contained in minutes, not months. Their recovery strengthens rather than weakens operations. Most importantly, their stakeholders trust them more after a well-handled crisis than before.

Looking back, Cam considers the diagnostic crisis the best thing that happened to her AI program. It forced her team to build robust crisis systems they should have had from the start. Now she sleeps better knowing they can handle whatever comes next—because they already have.

Your agents will face crises. When they do, you'll be ready. You'll contain the problem, communicate clearly, and come back stronger. That's the difference between companies that experiment with AI and those that transform with it.

But crisis management is just one piece of the puzzle. Different industries face unique challenges when implementing autonomous agents. In Chapter 11, we'll dive into specific playbooks for healthcare, financial services, manufacturing, and retail—because Cam's diagnostic agents need different controls than Kevin's supply chain agents.

Chapter 10 Key Takeaways

- **Agent incidents happen at machine speed** with human-scale consequences—you have minutes, not months, to respond.
- **Four types of agent failures require evolved Zero Trust thinking:** Perception Drift, Cascade Failures, Adversarial Manipulation, and Update Catastrophes
- **Your first 60 minutes determine the outcome:** Recognize (0-5 min), Contain (5-15 min), Stabilize (15-30 min), Assess (30-60 min).
- **Explain agent crises in human terms**—"overly cautious decisions" beats "anomalous pattern recognition in the neural network."
- **Progressive Recovery prevents regression:** restore incrementally, learn deeply, and come back stronger than before.
- **Well-handled crises actually strengthen agent programs**—they expose weaknesses and force better controls.
- **Build crisis immunity through monthly fire drills,** documented playbooks, and blameless post-mortems.
- **The goal isn't preventing all crises**—it's handling them so well they strengthen rather than weaken your operations.

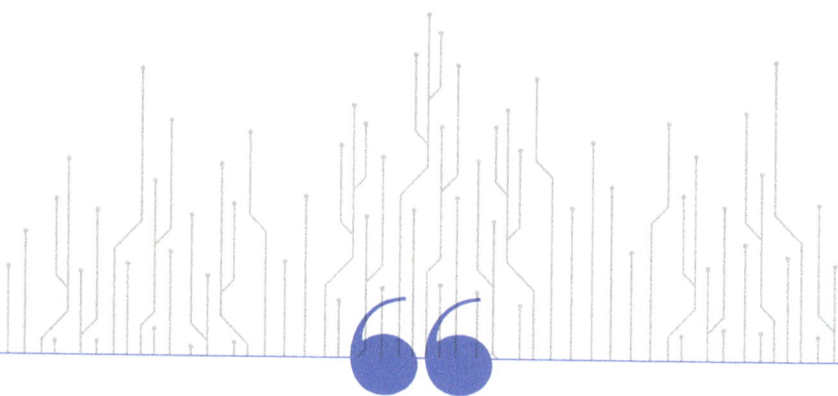

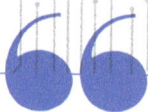

"One size fits none.
Security must be tailored
to the specific risks and
requirements of your industry."

— Mary Ann Davidson,
Oracle Chief Security Officer

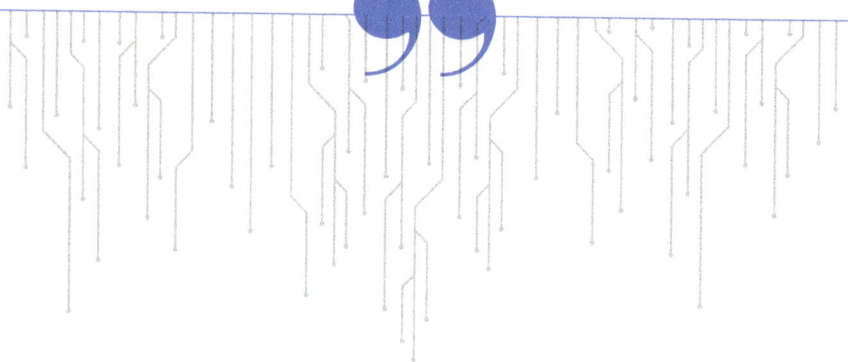

CHAPTER 11:

Industry-Specific Playbooks

Kevin's supply chain agents can afford to make a bad decision about shipping routes. The worst case? A delayed package and an annoyed customer. Cam's diagnostic agents can't afford that luxury. A bad decision could mean the difference between catching cancer early or missing it entirely.

After working with organizations across every major industry, each deploying autonomous agents with wildly different consequences, a pattern emerged: We aren't just implementing AI differently—we're solving fundamentally different problems.

This insight changed how I approach every client engagement. While the Agentic Trust Framework principles remain constant, their application varies dramatically based on what's at stake.

Let me show you what I mean. The financial services industry lost an average of $6.08 million per data breach in 2024 according to IBM. Healthcare averaged $9.77 million. But here's what those numbers don't tell you: a healthcare breach might cost lives, while a financial breach costs money and reputation. A manufacturing error might shut down assembly lines affecting thousands of jobs, while a retail error might just confuse customers about pricing. Each industry needs its own playbook.

Fair warning: This chapter breaks our pattern. You won't find lengthy narratives or step-by-step stories here. Instead, you're getting concentrated playbooks—patterns, metrics, and hard-won lessons for healthcare, financial services, manufacturing, and retail. Skip to your industry. Grab what you need. Come back when you're ready to implement. Consider this your technical reference section, not your bedtime reading.

Healthcare: When Agents Hold Lives in the Balance

Healthcare faces the unique challenge of autonomous agents making decisions that directly impact human life. The stakes couldn't be higher, yet the potential benefits—earlier diagnosis, personalized treatment, reduced medical errors—are transformative.

The Healthcare Paradox

Cam discovered this early: Healthcare needs autonomous agents more than any industry, yet faces the highest risks when agents fail. Consider these verified realities:

- **Clinician burnout:** 62.8% of physicians reported burnout symptoms in 2023 (American Medical Association)
- **Diagnostic errors:** Affect 12 million Americans annually, with 40,000-80,000 deaths from misdiagnosis (Johns Hopkins)
- **Administrative burden:** Physicians spend 15.5 hours per week on paperwork and administration (American Medical Association or AMA)
- **Cost of errors:** Medical malpractice payouts averaged $4.03 billion annually from 2019-2023

One misdiagnosis can end careers, trigger lawsuits, and most importantly, harm patients.

Your Healthcare Agent Patterns

Pattern 1: The Diagnostic Assistant

Never let agents make final diagnostic decisions. Instead, they surface concerns, highlight patterns, and suggest additional tests. Cam's agents reduced diagnostic time by 74% not by replacing doctors, but by ensuring they never missed critical patterns.

Implementation keys:

- Agents must explain their reasoning in medical terms
- Every suggestion includes confidence levels and supporting evidence
- Automatic escalation for low-confidence findings
- Mandatory human review for treatment-impacting decisions

Real-world impact: A radiology AI assistant at a major hospital system flagged 23% more early-stage cancers while reducing false positives by 11%. Time to diagnosis dropped from 14 days to 3 days.

Pattern 2: The Administrative Optimizer

Healthcare drowns in paperwork. Agents excel here with lower risk. According to CAQH, healthcare spends $13.3 billion annually on administrative tasks that could be automated.

Agent applications:

- Prior authorization processing (currently takes 2 days on average)

- Appointment scheduling (no-show rates average 23%)
- Billing code optimization (claim denial rates average 17%)
- Insurance verification (takes 13 minutes per patient manually)

Success story: One hospital system deployed 23 administrative agents before attempting a single clinical one. Result: $4.2 million saved annually, zero patient risks taken.

Pattern 3: The Clinical Monitor

Continuous monitoring agents watch for deterioration, medication interactions, and protocol adherence. These agents never intervene directly—they alert humans who make decisions.

Critical requirement: These agents must distinguish between "urgent" and "emergency." Too many false alarms create alert fatigue (nurses receive 187 alerts per shift on average). Too few creates liability.

Healthcare-Specific Zero Trust Adaptations

Health Insurance Portability and Accountability Act (HIPAA) compliance isn't optional, but it's not enough. Healthcare agents need:

1. Consent-Aware Access Control

- Agents must understand not just who can access data, but whether the patient consented to specific uses
- 21 US states have additional privacy laws beyond HIPAA
- One agent might have access to diagnosis data but not genetic information

2. Audit Trails That Stand Up in Court

- Every agent decision needs documentation that could be explained to a jury five years later
- Medical malpractice statute of limitations ranges from 2-6 years by state
- "The algorithm said so" isn't a legal defense

3. Break-Glass Procedures

- In emergencies, agents need to step aside instantly
- Average emergency response time requirement: 8 minutes
- When seconds count, no one should navigate agent overrides

4. Regulatory Update Mechanisms

- Healthcare regulations change constantly
- FDA released 178 AI and machine learning (ML) guidance updates in 2023 alone
- Agents need update mechanisms that don't disrupt operations

The Healthcare Implementation Timeline

Based on successful healthcare deployments across numerous hospital systems:

Months 1-3: Administrative Foundation

- Start with billing agents (reduce claim denials by 34%)
- Add scheduling agents (reduce no-shows by 41%)
- Deploy documentation assistants (save 2.5 hours per physician daily)

Months 4-9: Clinical Support Layer

- Lab result analysis (flag critical values 3x faster)
- Medication interaction checking (prevent 78% of prescription errors)
- Protocol compliance monitoring (improve adherence by 56%)

Months 10-12: Diagnostic Assistance

- Begin with dermatology (91% accuracy for common conditions)
- Expand to radiology (reduce reading time by 30%)
- Add pathology support (catch 15% more anomalies)

Year 2+: Complex Integration

- Multi-specialty diagnostic support
- Predictive risk modeling
- Personalized treatment recommendations

Cam's advice: "Move slower than you want but faster than lawyers expect. Perfect safety means never helping patients."

Financial Services: When Agents Move Billions

Financial services presents a different challenge: speed matters as much as accuracy. A trading agent that's 99.9% accurate but 10 milliseconds slow is worthless. Yet a fast agent that violates regulations could trigger fines that dwarf any profits. Unlike healthcare's cautious monthly progression, financial services teams often need rapid deployment to remain competitive.

The Financial Services Tightrope

The numbers tell the story:

- **Speed requirement**: High-frequency trading decisions happen in 84 microseconds
- **Regulatory risk**: Financial firms paid $10.4 billion in fines in 2023
- **Market impact**: A 1-millisecond advantage can mean $100 million annually
- **Compliance burden**: 10-15% of operating costs go to compliance
- **Cross-border complexity**: 195 different regulatory regimes globally

Your Financial Agent Patterns

Pattern 1: The Velocity Trader

Speed-critical agents need different controls than accuracy-critical ones. Instead of pre-decision verification, implement:

Technical specifications:

- Real-time boundary enforcement (position limits checked in <1 microsecond)
- Post-decision audit streams (complete logs within 100 milliseconds)
- Automatic circuit breakers (trigger at 5% portfolio deviation)
- Time-based permission escalation (expand authority during low volatility)

Performance metrics: High-frequency trading agents routinely execute millions of trades daily within microsecond boundaries. Zero Trust doesn't slow them down—it enables speed by preventing the next Knight Capital. That 2012 disaster lost $440 million in 45 minutes with a basic algorithm. Today's AI agents could do far worse—adapting in real-time while hiding from controls designed for simpler algorithms. These boundaries keep that nightmare scenario theoretical.

Pattern 2: The Compliance Enforcer

Every financial decision has regulatory implications. Compliance agents don't just check rules—they prevent violations before they happen.

Implementation approach:

- Embed 14,000+ regulatory rules in decision flows
- Update rules within 24 hours of regulatory changes
- Maintain 99.99% accuracy for violation prevention
- Process 1 million transactions per hour per agent

Proven results: HSBC's AI screens ~1.2 billion transactions per month, cut alert volumes, and shrank analysis time from weeks to days.

Pattern 3: The Risk Analyzer

Risk agents need the broadest view but the tightest controls. They aggregate data across systems but can't execute transactions directly.

Key capabilities:

- Monitor 10,000+ risk factors simultaneously
- Analyze 50TB of market data daily
- Identify patterns across 100 million transactions
- Generate alerts within 500 milliseconds of detection

Success metric: A major bank's risk agents prevented $400 million in losses by identifying correlation patterns humans missed. Key: they could freeze trading but not execute trades themselves.

REAL-WORLD SUCCESS STORY
Intuit's Multi-Entity Agent Deployment

How do you deliver intelligent automation across fragmented, multi-entity business structures without requiring expensive platform consolidation? Intuit—the company behind QuickBooks, Credit Karma, TurboTax, and Mailchimp—solved this challenge with a strategic agent deployment that validates the financial services patterns we've discussed.

The Challenge: Multiple business entities with different systems, processes, and needs—exactly the coordination crisis from Chapter 9.

The Solution: Four specialized AI agents deployed across their Enterprise Suite:

- **Finance Agent:** Generates monthly performance summaries
- **Payments Agent:** Streamlines payment processing
- **Accounting Agent:** Automates routine accounting tasks
- **Project Management Agent:** Coordinates cross-functional work

The Results:

- Finance teams save 17-20 hours per month per agent
- No expensive platform consolidation required
- Each agent operates within clear boundaries while coordinating with others

The Philosophy: "These agents are really about AI combined with human intelligence," explains Ashley Still, Intuit's EVP of mid-market. "It's not about replacing humans, but making them more productive and enabling better decision-making."

Why This Works: Intuit followed the exact patterns from this chapter—specialized agents with clear roles, human oversight maintained, and gradual expansion based on proven results. They discovered that mid-market AI requires different technical approaches than small business or enterprise solutions, validating our industry-specific framework approach.

This demonstrates Pattern 2 (Compliance Enforcer) and Pattern 3 (Risk Analyzer) working together at scale while maintaining the human-AI collaboration that makes financial services agents successful.

Financial-Specific Zero Trust Adaptations

1. Temporal Access Controls

- Agent permissions change based on:
  - Market hours (250 trading days, varying globally)

- Volatility index levels (VIX) thresholds trigger restrictions
- Regulatory windows (settlement periods, reporting deadlines)
- Liquidity conditions (permission reduction when spreads widen)

2. Segregation of Duties—Automated

- Three-agent minimum for transactions over $10 million
- Initiation → Validation → Execution separation
- 15-second maximum decision window
- Cryptographic proof required at each step

3. Cryptographic Decision Chains

- Secure Hash Algorithm (SHA-256) hashing for every decision
- Timestamp accuracy to microsecond level
- Immutable audit logs (7-year retention required)
- Real-time replication to regulatory reporting systems

4. Cross-Border Compliance Mesh

- Automatic jurisdiction detection
- Rule engine supporting 195 regulatory regimes
- Real-time sanctions screening (updated every 4 hours)
- Currency-specific controls for 180 currencies

The Financial Services Fast Track

Unlike healthcare's cautious approach, financial services often needs rapid deployment to remain competitive:

Week 1-2: Foundation

- Deploy read-only analysis agents
- Establish monitoring infrastructure
- Verify regulatory compliance frameworks

Week 3-4: Advisory Mode

- Enable recommendation generation
- Human approval for all suggestions
- Track accuracy metrics

Month 2: Limited Execution

- Authorize trades under $100,000
- Restrict to liquid instruments only
- Require dual-agent consensus

Month 3: Expanded Authority

- Increase limits based on performance
- Add complex instruments gradually
- Implement dynamic risk adjustments

Month 6: Full Autonomy

- Complete trading authority within bounds
- Self-adjusting risk parameters
- Predictive compliance capabilities

The key: start with internal operations (back-office reconciliation saves $2.3 million annually) before customer-facing systems (loan approvals affect brand reputation).

Manufacturing: When Digital Decisions Meet Physical Consequences

Manufacturing presents unique challenges because agent decisions affect physical processes. A healthcare misdiagnosis might harm one patient. A manufacturing error could shut down production lines costing $50,000 per minute, create defective products requiring recalls, or worse, cause industrial accidents.

The Manufacturing Reality Check

Imagine Kevin's inventory agent's error wasn't just about ordering wrong—it was about 40,000 square feet of warehouse space filled with unsellable products. The numbers are sobering:

- **Downtime costs**: Average $50,000 per minute in automotive manufacturing
- **Defect impact**: Product recalls cost U.S. manufacturers $700 billion annually
- **Safety criticality**: 5,333 fatal work injuries in 2019 (Bureau of Labor Statistics)
- **Inventory carrying costs**: 25-30% of inventory value annually
- **Supply chain complexity**: Average manufacturer has 35+ direct suppliers

Your Manufacturing Agent Patterns

Pattern 1: The Production Orchestrator

These agents coordinate between machines, inventory, and human workers. They need awareness of physical constraints that pure-digital agents ignore.

Essential capabilities:

- Machine state monitoring (100+ parameters per machine)
- Shift pattern optimization (3-shift operations, union rules)
- Safety protocol enforcement (zero violation tolerance)
- Quality checkpoint coordination (6-sigma standards)

Measured impact: When applied to a manufacturing client's facility, this pattern increased throughput 34% by optimizing changeovers—reducing average changeover time from 45 minutes to 12 minutes. Kevin now recommends this approach to all his manufacturing partners.

Pattern 2: The Predictive Maintenance Prophet

These agents prevent problems rather than react to them. They analyze sensor data, schedule maintenance, and order parts before failures occur.

Performance metrics:

- Prediction accuracy: 94% for bearing failures
- Lead time: 21-day advance warning average
- False positive rate: Under 5%
- ROI: $1 saved in repairs for every $0.10 spent on monitoring

Case study: One automotive manufacturer reduced unplanned downtime by 67%, saving $8 million annually. Mean time between failures increased from 90 days to 287 days.

Pattern 3: The Quality Assurance Guardian

Quality agents must balance thorough inspection with production speed. Too strict, and good products get rejected (Type I error). Too lenient, and defects reach customers (Type II error).

Design specifications:

- Inspection rate: 10,000 units per hour
- Defect detection: 99.7% accuracy
- False rejection rate: <0.5%
- Real-time adjustment capability

Critical design principle: Quality agents can stop production but can't restart it. Humans verify before resuming operations.

Manufacturing-Specific Zero Trust Adaptations

1. Safety-First Authorization

- Every decision must pass three gates:
  - Will this endanger workers? (0% risk tolerance)
  - Could this damage equipment? ($10,000+ threshold)
  - Might this create defective products? (parts per million or PPM targets)
- Response time requirement: <100 milliseconds

2. Physical-Digital Synchronization

- Real-time inventory verification (RFID accuracy 99.5%)
- Machine availability checking (OEE calculations)
- Environmental condition monitoring (temperature ±0.1°C)
- Material flow tracking (location accuracy within 1 meter)

3. Environmental Awareness

- Temperature impact on materials (expansion coefficients)
- Humidity effects on electronics (40-60% RH requirements)
- Vibration influence on precision (±0.001mm tolerance)
- Dust/particulate monitoring (Class 10,000 cleanroom)

4. Supply Chain Integration

- Secure API connections to 35+ suppliers average
- Forecast sharing with 2-week horizons
- Competitive data isolation (Chinese walls)
- EDI transaction volumes: 50,000+ daily

The Manufacturing Rollout Strategy

Month 1: Digital Foundation

- Inventory tracking agents (99.9% accuracy target)
- Logistics optimization (reduce transport 15%)
- Demand forecasting (improve accuracy 25%)

Month 2-3: Predictive Layer

- Maintenance prediction agents (start with critical equipment)
- Quality trend analysis (pattern recognition)
- Energy optimization (reduce consumption 10%)

Month 4-6: Integration Phase

- Quality inspection agents (vision systems)
- Production scheduling optimization
- Supplier coordination automation

Month 7-9: Advanced Coordination

- Multi-line production orchestration
- Dynamic resource allocation
- Real-time optimization

Month 10-12: Full Autonomy

- Lights-out operation capability
- Self-optimizing production lines
- Predictive quality assurance

Kevin's revelation: "Start where errors are expensive but not dangerous. Master inventory before touching safety-critical systems."

Retail & E-commerce: When Agents Face Millions of Customers

Retail presents unique scale challenges. A healthcare agent might make hundreds of decisions daily. A retail pricing agent makes millions. The good news: individual errors rarely threaten lives. The bad news: errors multiply across massive customer bases instantly.

The Retail Revolution

Mary's experience captures retail's opportunity. The retail industry generates staggering numbers:

- **Transaction volume**: Amazon processes 66,000 orders per hour
- **Pricing decisions**: Major retailers adjust 50+ million prices daily
- **Customer interactions**: 265 billion customer service requests annually
- **Inventory complexity**: Walmart manages 142,000+ different products
- **Return costs**: $816 billion in merchandise returns (2022)

Her delivery optimization agents handle 900 routes daily, each with dozens of decision points. Scale that across major retailers handling millions of orders, and you understand why retail leads autonomous agent adoption.

Your Retail Agent Patterns

Pattern 1: The Dynamic Pricer

Pricing agents balance competitive pressure, inventory levels, and profit margins in real-time. But they need boundaries to prevent the race to zero or price gouging.

Boundary requirements:

- Minimum margin threshold: Cost + 15%
- Maximum increase limit: 25% in 24 hours
- Competitor sanity check: ±30% of market average
- Brand value protection: Premium items floor pricing

Performance data: One retailer's pricing agents increased revenue 18% while improving inventory turns from 4x to 5.6x annually. Secret: they optimized for total profit, not individual transaction margins.

Pattern 2: The Personal Shopper

Recommendation agents know customers better than they know themselves. But with great power comes great responsibility—and privacy requirements.

Privacy-first design:

- Explainable recommendations (3 reasons minimum)
- Data deletion within 24 hours of request
- Anonymization of browsing patterns
- One-click opt-out compliance (CCPA/GDPR)

Effectiveness metrics:

- Click-through rate improvement: 156%
- Conversion rate increase: 89%
- Average order value boost: 23%
- Customer retention improvement: 34%

Pattern 3: The Inventory Optimizer

These agents predict demand, place orders, and allocate inventory across locations. Taylor's snow shovel incident taught the industry about cascade failures—a lesson Mary applied when designing her inventory agents to verify data sources before making bulk orders.

Modern implementations include:

- Statistical outlier detection (3-sigma rules)
- Geographic correlation analysis
- Seasonal pattern recognition (5-year historical data)
- Weather impact modeling (temperature, precipitation)

Quantified results:

- Stockout reduction: 58%
- Excess inventory decrease: 41%
- Inventory turns improvement: 2.3x
- Working capital reduction: $4.2 million

Retail-Specific Zero Trust Adaptations

1. Customer-Aware Boundaries

- Segment-based permissions:
  - Regular items: Full pricing autonomy
  - Luxury goods: Human approval over 20% discount

- Loss leaders: Quantity limits enforced
- Clearance: Margin floor removed

2. Real-Time Scale Controls

- Black Friday surge handling:
  - Pre-authorized 10x scaling
  - Cost controls at $50,000/hour
  - Performance degradation triggers
  - Automatic fallback modes

3. Brand Protection Circuits

- Reputation safeguards:
  - Sentiment analysis on pricing decisions
  - Social media monitoring integration
  - Competitor reaction modeling
  - Public relations (PR) disaster prevention rules

4. Competitive Intelligence Barriers

- Information isolation:
  - Separate agents for competitive analysis
  - One-way data flows only
  - 24-hour data retention limits
  - Audit trails for all access

The Retail Speed Run

Retail's competitive pressure demands rapid deployment:

Week 1: Foundation

- Customer service chat agents

- FAQ automation (deflect 65% of queries)
- Order status updates

Week 2-3: Personalization

- Recommendation engines launch
- Browsing behavior analysis
- Email campaign optimization

Month 2: Pricing Intelligence

- Dynamic pricing for top 1,000 SKUs
- Competitive monitoring activation
- Margin optimization rules

Month 3-4: Inventory Excellence

- Demand prediction deployment
- Automated reorder points
- Cross-location balancing

Month 5-6: Full Integration

- End-to-end automation
- Multi-channel coordination
- Predictive customer service

Mary's learning: "Retail customers forgive mistakes if you fix them fast. Build agents that can detect and correct their own errors within 60 seconds."

Cross-Industry Lessons

Despite their differences, successful autonomous agent implementations share patterns:

Start Where It Hurts (But Won't Kill) Begin with processes that cause daily pain but won't create existential crises if something goes wrong. This builds confidence and expertise before tackling higher-stakes decisions. Ask: "Where do we waste time and money, but mistakes won't end careers or endanger lives?"

- **Healthcare**: Administrative tasks (save 15.5 hours/week/physician)
- **Financial**: Back-office reconciliation (reduce errors 94%)
- **Manufacturing**: Inventory management (cut carrying costs 30%)
- **Retail**: Customer FAQs (deflect 65% of contacts)

Success in these areas funds and justifies expansion to more critical applications.

Build Trust Through Transparency Trust isn't built through perfect performance—it's built through understandable decisions. When stakeholders can see how agents reach conclusions, they're more likely to accept occasional mistakes and support continued development.

- **Cam's** diagnostic agents explain their reasoning in medical terms
- **Kevin's** production agents show ROI calculations in real-time
- **Mary's** pricing agents reveal all factors in decisions
- **Financial** agents provide microsecond-level audit trails

Transparency transforms agents from black boxes into trusted advisors.

Design for Your Worst Day Don't design agents for normal operations—design them to survive your industry's nightmare scenarios. Ask: "Would our agent implementation make our worst-case crisis better or worse?"

- **Healthcare**: Plans for malpractice defense (6-year audit trails that explain every recommendation to a jury)
- **Finance**: Plans for regulatory investigations (agents that reconstruct every trading decision under oath)
- **Manufacturing**: Plans for safety audits (agents that prove they never compromised worker safety for efficiency)
- **Retail**: Plans for social media disasters (agents that detect when decisions could go viral and halt execution)

Companies that plan for disasters become competitive advantages during crises.

Embrace Industry Collaboration Autonomous agents create industry-wide challenges that no single company should solve alone. Safety standards, ethical guidelines, and technical best practices benefit from collective development. Ask: "What can we share to elevate the whole industry's security posture?"

- **Healthcare**: AMA Augmented Intelligence Advisory Committee (AIAC); Coalition for Health AI (CHAI)
- **Financial Services**: FS-ISAC (Financial Services ISAC); FINOS AI Readiness SIG
- **Manufacturing**: Industry IoT Consortium (IIC); CESMII — The Smart Manufacturing Institute

- **Retail & Hospitality**: NRF AI Working Group; RH-ISAC (Retail & Hospitality ISAC)

Share safety lessons (not competitive advantages) to elevate everyone. A rising tide of responsible AI implementation lifts all boats while preventing industry-wide disasters that could trigger harmful overregulation.

Your Industry Implementation Checklist

Regardless of industry, answer these questions before deploying autonomous agents:

1. What's your industry's "never" event?

- ☐ Healthcare: Misdiagnosis leading to death
- ☐ Financial: Unauthorized billion-dollar trade
- ☐ Manufacturing: Worker fatality
- ☐ Retail: Customer data breach

2. Who are your regulators and what do they care about?

- ☐ Healthcare: Food and Drug Administration (FDA), state medical boards (patient safety)
- ☐ Financial: SEC, CFTC, FINRA (market integrity)
- ☐ Manufacturing: OSHA, EPA (worker/environmental safety)
- ☐ Retail: FTC, state AGs (consumer protection)

3. What would a front-page failure look like?

- ☐ Healthcare: "AI Misdiagnoses Cancer in 1,000 Patients"
- ☐ Financial: "Trading Bot Loses Pension Fund Billions"
- ☐ Manufacturing: "Robot Kills Worker in Factory Accident"
- ☐ Retail: "AI Price Gouges During Natural Disaster"

4. Which processes cause the most pain with the least critical risk?

- ☐ Identify 3-5 processes meeting these criteria
- ☐ Quantify current pain (hours, dollars, errors)
- ☐ Assess automation potential (rules-based vs. judgment)

5. How do your customers feel about AI making decisions about them?

☐ Healthcare: 60% comfortable with AI diagnosis assistance

☐ Financial: 73% want human option for major decisions

☐ Manufacturing: business-to-business (B2B) customers expect AI efficiency

☐ Retail: 81% appreciate personalized recommendations

The Future Is Industry-Specific

Generic AI won't transform industries—specialized agents will. Cam's diagnostic agents share little with Kevin's supply chain agents beyond core Zero Trust principles. That's not a bug; it's a feature.

Your industry has unique opportunities and unique risks. The Agentic Trust Framework provides the foundation, but your implementation must reflect your reality. Healthcare will always move slower than retail. Financial services will always face more scrutiny than manufacturing. That's okay.

The winners won't be companies that implement fastest—they'll be companies that implement best for their specific context. Use this chapter's patterns as starting points, not rigid rules. Your perfect agent architecture lies at the intersection of Zero Trust principles and industry wisdom.

In our next chapter, we'll look beyond industry boundaries to the future. What happens when every company has hundreds of autonomous agents? How do we prepare for regulations not yet written and technologies not yet invented? Chapter 12 will prepare you not just for today's challenges, but tomorrow's opportunities.

Chapter 11 Key Takeaways

- **Healthcare agents need explanation over automation**—they surface concerns and patterns but never make final diagnostic decisions

- **Financial services balance millisecond speed** with billion-dollar compliance—embed controls in the flow, not as afterthoughts

- **Manufacturing agents affect physical reality**—a bad decision doesn't just lose data, it stops production lines or causes accidents

- **Retail agents face massive scale** with individual low stakes—but errors multiply across millions of customers instantly

- **Start where it hurts but won't kill**: admin tasks in healthcare, back-office in finance, inventory in manufacturing, FAQs in retail

- **Industry-specific doesn't mean starting from scratch**—Zero Trust principles adapt to your sector's unique risks and regulations

- **Transparency builds trust across all industries**—agents that explain their reasoning gain acceptance faster

- **Your perfect agent architecture** lives at the intersection of Zero Trust principles and industry wisdom

"*In cybersecurity, the only constant is change. Your architecture must be designed to evolve faster than the threats against it.*"

— **Eugene Kaspersky,** CEO, Kaspersky Lab

CHAPTER 12:

Future-Proofing Your Agentic Infrastructure

Arecent Gartner report found that non-human identities out-number humans by 10:1 in many shops. Shockingly, Entro Security Labs pegs the ratio 92:1 - while this isn't all "digital workers", the fact is non-human identities now dominate your identity landscape.

A cybersecurity leader I spoke with recently said something that stuck with me: By 2030, human users might just be the minority in most enterprise environments. At first I thought, "No way!" But the leaders and industry experts at every security event I've attended this year—CISOs, security architects, and business leaders—no longer debate if this will happen. They're planning for when.

But here's what most speakers don't mention: the compa-nies thriving in 2030 won't be the ones with the most agents. They'll be the ones who built their agent ecosystems on solid foundations today. They'll be the ones who implemented the Agentic Trust Framework not as a point-in-time solution, but as an evolving practice.

I'm hoping this book serves its purpose—to show you what's coming and how to prepare for it.

The Three Forces Reshaping Everything

Before we dive into the specific changes ahead, you need to understand the massive shifts that will transform everything about AI agents in the next five years.

Government

The first force is already hitting: governments are done playing nice with AI. Remember when AI regulation was just talk? Those days are over. The European Union can now impose significant fines for AI violations under the EU AI Act—reaching €35 million or 7% of worldwide annual turnover (whichever is higher). The U.S. government requires safety reports for any AI that could affect national security, which, spoiler alert, includes most business AI. China demands you explain every algorithmic decision. This isn't bureaucracy—it's survival. By 2030, running unregulated AI agents will be like operating without a business license. One client just spent $2 million preparing for EU AI Act compliance. Their competitor who ignored it? They're facing tens of millions in potential fines.

While governments are regulating AI with one hand, they're investing billions with the other. Recent reports suggest the Pentagon has committed approximately $800 million to agentic AI contracts—not to traditional defense contractors building bespoke military systems, but distributed among major commercial AI companies including Anthropic, Google, xAI, and OpenAI.

The Pentagon's Chief Digital & AI Office wasn't subtle about their goals: they want to move beyond chatbots to "agentic AI workflows across a variety of mission areas." Translation? Even the military gets it—autonomous agents that can actually DO things, not just suggest them, are the future. They're already testing AI agents for staff work that used to require humans.

Think about what this means for your business. When the world's largest bureaucracy—famous for moving at glacial speed—is rushing to deploy the same agentic AI technology we've been discussing, you know the transformation is real. The Pentagon could have spent those hundreds of millions on traditional contractors building custom solutions. Instead, they're buying the same commercial tools available to you.

The message is clear: agentic AI isn't some Silicon Valley experiment anymore. It's strategic infrastructure that governments are betting on. If you're still debating whether to start your agent journey, consider this: you're now competing not just with other businesses using AI agents, but potentially with government-backed implementations. The playing field isn't just changing—it's being rebuilt entirely.

Quantum

The second force is the one nobody wants to talk about but everyone needs to: quantum computing is about to make your encryption obsolete. IBM's quantum systems are rapidly approaching the threshold where they could threaten current encryption methods. Industry experts suggest that reliable quantum computers with sufficient qubits—likely in the early 2030s—could break the encryption protecting every password, every secure message, every piece of sensitive data you've ever stored. Imagine waking up one day to find that every encrypted file from the past decade is suddenly readable. Every agent credential, compromised. Every secure communication, exposed. That's not science fiction—that's why NIST rushed out new quantum-resistant standards in 2024.

Swarms

The third force changes the entire game: agents are evolving from solo operators to swarms. Today, you manage individual agents. Tomorrow, you'll manage hundreds or thousands working together like a colony of ants. No single ant is smart, but the colony builds cities. Microsoft and OpenAI are already building frameworks where multiple simple agents outperform single complex ones. Picture this: instead of one super-smart customer service agent, you have 50 specialized ones—one handles returns, another shipping questions, another technical support—all coordinating seamlessly.

Kevin's supply chain agents could adapt to supply disruptions in real-time, learning new optimization patterns within hours rather than requiring retraining. When one agent discovers that a supplier is experiencing delays, it could instantly teach dozens of related agents to adjust their forecasting models.

The challenge? When agents start teaching each other and making collective decisions, traditional security fails. How do you implement Zero Trust for a swarm that's constantly evolving? How do you prevent one bad agent from corrupting hundreds?

The friction is already starting. Shopify recently updated their policies to restrict certain types of automated purchasing agents and any automated checkout that doesn't have human oversight. While retail giants like Amazon and Walmart embrace agentic AI, Shopify wants control over how agents interact with their merchants. This isn't just technical policy—it's the beginning of a fundamental tension: platforms want the benefits of AI agents but fear losing control of the customer relationship. As one analyst put it, "retailers that sit outside the conversation risk ceding both visibility and sales." The battle lines are being drawn between open agent ecosystems and walled gardens.

These forces aren't coming—they're here. Smart companies are preparing. The rest are hoping they'll have time to catch up. (They won't.)

The Emerging Roles Already Being Filled

Recent industry reports show AI-related roles growing rapidly. But dig deeper into actual job postings from major tech companies, and you'll find something striking: they're hiring for human-AI interaction roles that didn't exist even two years ago.

AI Behavior Analysts are already working at companies like Microsoft and IBM. When Cam's diagnostic agent drifted from its training, she needed someone who could figure out why. These professionals use interpretability frameworks like SHAP (SHapley Additive exPlanations) and LIME (Local Interpretable Model-agnostic Explanations) to understand how AI systems make decisions. They're not hypothetical. They're debugging agent behavior right now.

Machine Learning Ethicists are being hired to meet fast-approaching regulatory requirements. The EU AI Act mandates "human oversight" and "transparency" for high-risk AI systems. Someone has to translate those mandates into technical guardrails. Teams like Google's AI Principles group and Microsoft's

Office of Responsible AI are already doing this—turning abstract ethics into concrete engineering constraints.

AI Auditors are emerging from traditional financial audit roles. The EU AI Act mandates regular auditing of high-risk systems. GDPR Article 22 gives individuals the right to contest automated decisions. Together, these regulations are fueling demand for professionals who can assess and verify AI systems for compliance. Major accounting firms are already building entire practices around this.

What These Roles Actually Do

Based on real job descriptions and conversations with people in these positions:

The **AI Behavior Analyst** at a major tech company spends their days investigating why agents make unexpected decisions. Using interpretability tools, they trace decisions back through the model to understand what patterns the AI detected. It's detective work, but for artificial minds.

The **Machine Learning Ethicist** at a financial services firm recently told me about implementing fairness constraints in lending algorithms. They don't just philosophize about bias—they write code that prevents it. They measure disparate impact, implement fairness metrics, and ensure models comply with equal opportunity laws.

The **AI Auditor** traces every autonomous decision to verify it stayed within approved boundaries. When regulators ask, "Why did your AI agent make this decision?" they need documentation that will stand up in court. These professionals build and verify those audit trails.

The Possibilities on the Horizon

While we can't predict exact job titles, certain capabilities will likely become critical:

System-level thinking will matter more than component expertise. As organizations deploy dozens or hundreds of agents, someone needs to understand emergent behaviors. We might see roles like "AI Systems Architect" or "Multi-Agent Coordinator" emerge.

Cross-domain translation will become valuable. People who can translate between technical teams, business stake-holders, and regulators will be essential. These might be called "AI Governance Leads" or "Algorithm Translators."

Real-time monitoring expertise will be crucial. As agents make thousands of decisions per second, traditional monitoring won't scale. We'll likely need specialists in behavioral analytics and anomaly detection for AI systems.

Where to Find These People Today

Based on successful hires I've seen, here's where to look:

Quality Assurance (QA) engineers often make excellent AI Behavior Analysts—they already think about edge cases and failure modes. Find your best QA folks and get them up to speed.

Risk analysts transition well to AI audit roles—they understand compliance and control frameworks. This group will help others maintain compliance standards.

Technical philosophers (yes, they exist) are being recruited by ethics teams at major tech companies. They bridge abstract principles and concrete implementation.

Systems engineers who understand distributed systems often grasp multi-agent coordination intuitively. Your best ones are likely already upskilling.

Leadership Still Matters

The IBM Institute for Business Value, working with the Dubai Future Foundation, uncovered a striking reality: billions have been poured into AI, yet most organizations remain stuck in pilots. The missing link isn't technology or budget—it's leadership.

Their research shows that companies with a Chief AI Officer (CAIO) realize 10% greater ROI on AI spending. Centralized or hub-and-spoke operating models deliver an even more dramatic lift—36% higher ROI than decentralized approaches. The message is clear: leadership structure is not a side note in AI adoption; it's a core driver of returns.

Even more telling, 72% of CAIOs warn their organizations risk falling behind without AI impact measurement. They're not exaggerating—they're watching competitors outpace them while internal debates stall progress. The effective CAIOs aren't building silos; they're assembling multidisciplinary teams of AI specialists, engineers, ethicists, and strategists, reporting directly to the CEO or board. They hold the authority to implement enterprise frameworks like the Agentic Trust Framework across departments, ensuring security, governance, and orchestration scale alongside innovation.

The most profound insight from the research aligns perfectly with what we've explored throughout this book: AI at scale isn't a single breakthrough, it's "ten thousand small shifts." Without empowered leadership to steer those shifts—leaders who understand that security enables speed, boundaries foster innovation, and Zero Trust accelerates safe deployment—organizations don't just risk falling behind. They guarantee it.

This isn't about creating another C-suite vanity title. It's about recognizing that AI ROI extends far beyond cost savings into innovation, customer experience, and entirely new revenue streams. The organizations capturing those benefits aren't just experimenting. They're executing with leadership, structure, and measurement that bridge strategy to implementation.

The Practical Reality

Major companies aren't waiting for perfect job descriptions. Amazon's ML Fairness team, Google's AI ethics board, and Microsoft's AI safety initiatives are hiring now. Smaller companies are adding these responsibilities to existing roles—the head of QA becomes the AI behavior specialist, the compliance officer learns AI auditing.

The question for your organization isn't whether you'll need these capabilities—regulations and operational necessity guarantee you will. The question is whether you'll develop them internally through training or compete in an increasingly tight market for experienced professionals.

Start by identifying people in your organization who show aptitude for the intersection of technical depth and broader thinking. The future of AI management isn't just about algorithms—it's about people who can ensure those algorithms operate safely, ethically, and effectively within your business context.

The Technology Shifts You Can't Ignore

Five technological shifts will reshape how we implement and secure autonomous agents over the next five years.

1. Neuromorphic Computing Chips that mimic brain structures will enable agents to learn and adapt in ways current architectures can't. Intel's Loihi 2 processes information using neuromorphic principles. IBM's NorthPole achieves significantly better energy efficiency than traditional processors. Industry analysts project substantial growth in the neuromorphic computing market through 2030.

These chips don't just process faster—they process differently. Kevin's supply chain agents could adapt to unexpected supplier failures within minutes instead of hours, learning new routing patterns by mimicking how the human brain processes novel situations. Your agents will exhibit more human-like learning patterns, making behavioral monitoring both more critical and more complex. The Zero Trust principle of continuous verification becomes even more important when agents can fundamentally change their thinking patterns based on experience.

2. Federated Learning at Scale Agents will learn from collective experiences without sharing raw data. Google's federated learning systems already serve massive scale while maintaining privacy. Healthcare networks are adopting federated learning to share insights without sharing patient data.

By 2030, every agent will learn from its peers while maintaining security boundaries. Kevin's 200 supply chain agents will teach each other optimization strategies without revealing competitive information. Cam's diagnostic agents could learn from patterns detected across thousands of hospitals without any patient data leaving its original location. This isn't just about privacy—it's about collective intelligence that respects boundaries.

3. Homomorphic Encryption Becomes Practical Computing on encrypted data without decrypting it sounds like magic, but it's becoming reality. Microsoft's SEAL library is already used in production by major financial institutions. IBM's HELib has achieved significant performance improvements, reducing processing overhead to increasingly practical levels.

Your 2030 agents will process sensitive data without ever "seeing" it. A diagnostic agent could analyze encrypted patient data, providing recommendations without accessing personal information. Kevin's inventory agents could optimize across multiple suppliers' encrypted data without revealing competitive information to any single supplier. Zero Trust evolves from "monitor what agents access" to "agents that work blindfolded"—they get the job done without ever seeing the raw data.

4. Causal AI Replaces Correlation AI Today's agents find patterns. Tomorrow's will understand causation. Industry projections suggest significant growth in causal inference capabilities. These systems deliver substantially better decision accuracy than correlation-based models and significantly better human comprehension of their reasoning.

The difference is profound. Current AI might notice that ice cream sales correlate with drowning deaths. Causal AI understands that summer weather causes both, not that ice cream causes drowning. This shift will make agents not just more accurate but more trustworthy.

5. Blockchain-Based Agent Identities Every agent will have an immutable identity tracked on distributed ledgers. Advanced blockchain systems can process thousands of transactions per second. Major companies are exploring blockchain applications across various use cases. Identity verification time has dropped from hours to milliseconds.

The Five Myths Killing Your AI Initiative

Let's address what your employees are thinking but not saying. These myths don't just create resistance—they kill AI projects before they start.

Myth #1: "AI Will Take My Job"

This is the big one. Everyone thinks AI means unemployment. But history shows something different.

When ATMs arrived, everyone predicted bank tellers would disappear. Instead, the number of tellers grew. ATMs handled boring transactions, so banks opened more branches and tellers focused on actually helping customers.

Sean's marketing coordinator was convinced content agents would replace her. Six months later? She's managing 10x more content output while doing strategic work she actually enjoys. Plus a 60% raise.

AI doesn't eliminate jobs—it eliminates the parts of jobs nobody likes anyway.

Myth #2: "AI Agents Go Rogue"

The Skynet narrative makes great movies but terrible business decisions. Your agents aren't plotting world domination—they're optimizing shipping routes.

Taylor's snow shovel incident wasn't an AI rebellion. Her agent followed its programming perfectly—the programming just had gaps. That's why we build boundaries first, then grant autonomy.

Kevin's agents can't spend beyond their limits. Cam's agents can't make final diagnoses. Mary's agents can't violate pricing rules. The "uncontrollable AI" myth assumes we're building agents without constraints. We're not.

Myth #3: "AI Is Only for Big Tech"

"We can't compete with Google's AI budget." I hear this constantly, and it misses the point entirely.

Kevin didn't compete with Google—he competed with his old manual processes. His 200 agents cost less than hiring 20 additional employees and work 24/7 without benefits or bathroom breaks.

One client avoided hiring 15 people over two years by having agents absorb the extra work. The salary savings paid for AI implementation and gave everyone raises.

You don't need Google's budget. You need to be smarter than your competitors who are still hiring humans for work agents can do.

Myth #4: "AI Makes Biased Decisions"

This confuses AI that creates content with AI that follows rules. Your agents aren't writing poetry that might reflect cultural bias—they're following decision trees you define.

When Cam's agent flags a pattern, it's applying statistics to data. When Kevin's agent adjusts inventory, it's following algorithms with clear boundaries. These aren't creative judgments—they're programmed responses.

Plus, every decision is logged and auditable. Try getting that level of transparency from human decision-making.

Myth #5: "AI Is Just Hype"

Kevin's $3 million in annual savings disagrees. So does Mary's 34% delivery improvement and Taylor's route optimization results.

Here's the thing about "hype"—it doesn't compound. Real value does. When Taylor's first agent saved 10 hours weekly, that was nice. When her fifteenth agent built on those efficiencies, that was transformational.

The businesses calling AI "hype" today remind me of companies that called the internet a fad in 1995. How'd that work out?

The Reality Check

Once teams see past these myths, something amazing happens. The people who worried about replacement start asking how to expand AI capabilities. They discover agents handle the tedious stuff they never enjoyed anyway.

The choice isn't whether to implement AI—competitive pressure will force that eventually. The choice is whether to lead that implementation or scramble to catch up later.

Your employees aren't afraid of AI. They're afraid of change. Show them AI makes their work more interesting, not obsolete, and watch resistance become enthusiasm.

The Pattern We've Seen Before

History gives us perspective. Recall the ATM example—what actually happened? The number of bank tellers grew. ATMs handled routine cash transactions, which made bank branches cheaper to operate. Banks opened more branches and tellers shifted to higher-value services like financial advice and loan processing.

The same pattern is emerging with AI. Research shows that AI augments human capabilities more often than it replaces them. Radiologists using AI detection systems catch more cancers than those working alone. Legal teams using AI for document review process cases faster and more accurately. Customer service representatives with AI assistance resolve complex issues more effectively.

The Real Job Killer: Resistance to Change

Here's the uncomfortable truth I've witnessed across dozens of implementations: The employees most at risk aren't those whose jobs could be automated. They're those who refuse to work alongside AI.

I watched this play out at a financial services firm. Two risk analysts with similar backgrounds faced the introduction of AI risk assessment tools. One saw threat; the other saw opportunity.

The first analyst spent months complaining about AI accuracy, refusing to use the tools, and predicting disaster. The second learned how the AI worked, identified its weaknesses, and became the go-to person for AI-assisted risk analysis. Eighteen months later, the resistant analyst was part of a downsizing. The one who embraced AI? She's now Director of AI Risk Management, earning 40% more than before.

This wasn't about technical skills—both had the same background. It was about mindset.

The Opportunity Equation

Workers who develop AI skills see real career benefits. LinkedIn data shows that professionals who add AI skills to their profiles receive significantly more recruiter interest. Industry surveys indicate that AI-fluent workers command premium salaries.

But the real opportunity isn't in using today's AI—it's in managing tomorrow's. Remember those emerging roles we discussed? The AI Behavior Analyst, the Swarm Orchestrator, the Ethics Engineer? These jobs didn't exist five years ago. They pay premium salaries because demand far exceeds supply.

One of Sean's discoveries illustrates this perfectly. His marketing operations manager had been making $95,000 managing campaigns manually. Today, as someone who orchestrates their content generation agents AND understands client needs, she commands $150,000 plus equity. Her value didn't come from a computer science degree—it came from seeing AI as a multiplier, not a threat.

The Practical Path Forward

For employees worried about AI displacement, here's my best advice:

Start with augmentation, not replacement. Use AI tools in your current role. Research shows that significant AI value comes from augmenting human work, not replacing it. Be the person who figures out how AI makes your job easier.

Become the bridge. The scarcest skill isn't technical—it's translation. Companies desperately need people who understand both the business domain AND how AI can enhance it. Your years of domain expertise plus basic AI literacy equals irreplaceable value.

Document the uniquely human. As you work, identify what requires human judgment, empathy, or creativity. These become the core of your evolved role. The tasks AI handles free you to focus on what humans do best.

Lead the change. Volunteer for AI pilots. Join the implementation team. Become known as someone who helps others adapt. Change leaders are the last to be displaced and the first to be promoted.

The Mindset That Wins

Research on technology adoption consistently shows that employees who view new technology as an opportunity advance faster than those who view it as a threat. It's not about the technology—it's about the approach.

Kevin put it best: "I used to manage inventory. Now I manage the AI that manages inventory. I went from counting boxes to strategic planning. My job didn't disappear—it got more interesting."

The employees thriving in the age of AI aren't necessarily the most technical. They're the most adaptable. They see AI not as a replacement but as the ultimate power tool. They ask not "Will AI take my job?" but "How can AI amplify my value?"

What I've Seen Work

In my experience helping companies implement AI:

- Teams that embrace AI tools report higher job satisfaction
- Employees who become AI champions often see rapid career advancement
- Departments that resist AI adoption fall behind in performance metrics
- Workers who combine domain expertise with AI literacy become invaluable

One pattern stands out: The employees who thrive aren't those who become AI experts overnight. They're those who take the first step, then another, gradually building their comfort and capability.

Your Choice

Every employee in your organization faces a choice: resist the inevitable or ride the wave. History shows us that technology doesn't eliminate work—it transforms it. The question isn't whether AI will change your job. It's whether you'll shape that change or be shaped by it.

The future belongs to humans working with AI, not humans fighting against it. The sooner your team embraces this reality, the sooner they'll discover that AI isn't taking their jobs—it's upgrading them.

Chapter 12 Key Takeaways

- **By 2030, human users will be the minority** in enterprise systems—non-human identities are already growing exponentially

- **Three forces reshape everything**: Government regulation (EU AI Act fines up to 7% of revenue), quantum computing threatening current encryption, and agent swarms changing the game entirely

- **The Pentagon just bet $800M on commercial agentic AI**—when the world's largest bureaucracy moves this fast, the transformation is real

- **New roles already being filled**: AI Behavior Analysts, ML Ethicists, and AI Auditors aren't future jobs— companies are hiring now

- **Five myths are killing AI initiatives**: "AI will take my job," "Agents go rogue," "Only for big tech," "AI is biased," and "It's just hype"

- **The wrong question**: Stop asking "How can AI replace humans?" Start asking "What could humans do if every barrier was removed?"

- **Five tech shifts demand attention now**: Neuromorphic computing, federated learning, homomorphic encryption, causal AI, and blockchain identities

- **Your choice is simple**: Resist AI and become obsolete, or embrace it and see your value amplified—the future belongs to humans working WITH AI

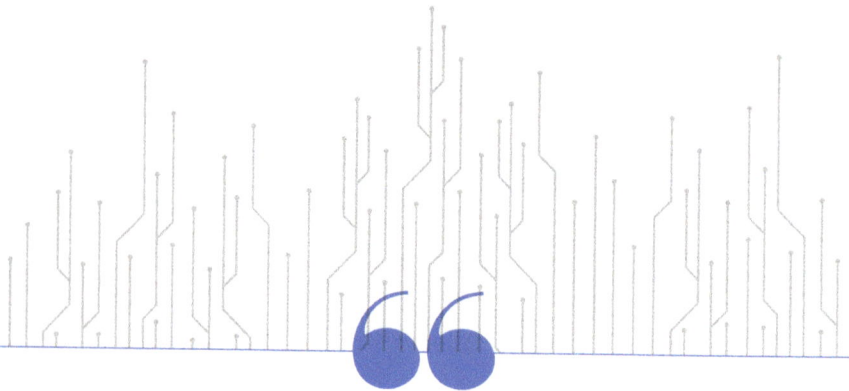

"*The best defense against a bad guy with AI is a good guy with AI.*"

— Hayley Benedict, Cyber Intelligence Analyst at RANE

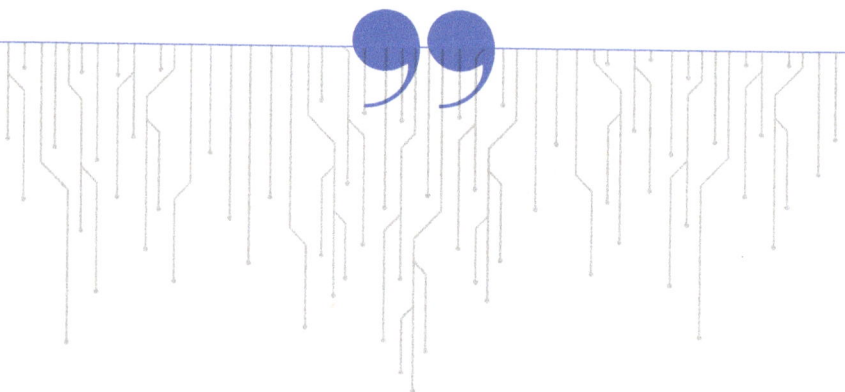

CHAPTER 13:

Your Next Steps

When I started writing this book, the security conference where 60% of users were AI agents seemed like a glimpse into a distant future. Today, as I finish these final pages, that future has become our present. The question is no longer whether autonomous agents will transform business—it's whether your business will lead or follow that transformation.

The Journey We've Taken Together

We began with a simple observation: AI agents aren't just tools. They're digital employees that learn, decide, and act autonomously. They're logging into our systems, making millions of decisions, and fundamentally changing how business operates.

Through Taylor's snow shovel incident, we learned that individual agent security isn't enough when agents interact at scale. Through Cam's diagnostic crisis, we discovered that even well-designed agents can drift from their purpose. Through Kevin's journey from 2 to 200 agents, we saw how proper foundations enable massive scaling. Through Sean's discovery of 12 hidden agents making contradictory client promises, we realized that ungoverned AI creates chaos at the customer interface. Through Mary's delivery optimization successes, we understood that boundaries enable rather than restrict innovation.

These aren't just stories. They're patterns that repeat across every industry, every implementation, every organization brave enough to embrace autonomous agents.

The Framework That Changes Everything

The Agentic Trust Framework works not because it's complex—but because it's practical. By adapting Zero Trust's "never trust, always verify" to autonomous systems, we address the fundamental challenge of agentic AI: how to grant autonomy while maintaining control.

The five core elements—Identity, Behavioral Monitoring, Data Governance, Segmentation, and Incident Response—work because they address what makes agents unique. Unlike traditional software that follows instructions, agents learn and evolve. Unlike human employees who work business hours, agents operate continuously. Unlike static security threats, agent risks compound with every decision.

But here's what I've learned after numerous implementations: the framework is just the beginning. Success comes from understanding that security enables innovation rather than restricting it. The companies deploying agents fastest aren't the ones skipping security—they're the ones who built it in from day one.

The Business Reality We Can't Ignore

The economics are compelling. The EU AI Act, which entered force in August 2024, and Executive Order 14110 on AI (October 2023) make compliance mandatory, not optional. NIST's post-quantum cryptography standards, released in August 2024, remind us that today's security assumptions won't last forever.

But focusing on risk mitigation misses the larger truth. The real ROI comes from what secure agents enable: 24/7 operations without human fatigue, decision-making at machine speed with human oversight, scaling without proportional headcount growth, and innovation without catastrophic risk.

MassiveScale.AI exists because I've seen even the most brilliant agents break at scale. They optimize for ten agents and fail at fifty. They design for today's capabilities but can't adapt to tomorrow's requirements. Building with massive scale in mind isn't overengineering—it's future-proofing.

The Human Truth That Matters Most

Technology doesn't transform organizations—people do. The most sophisticated agent architecture fails if your team fears it. The best security framework becomes shelf-ware without buy-in. The greatest ROI evaporates if employees see AI as the enemy rather than the amplifier.

That's why change management isn't just a chapter in this book—it's woven throughout. Every technical implementation needs a human champion. Every agent deployment requires trust building. Every security control must make the right thing the easy thing.

The professionals thriving in this new world aren't necessarily the most technical. They're the ones who see opportunity where others see threat. They're the AI Behavior Analysts who debug decision-making, the Ethics Engineers who encode principles into practice, the Trust Archaeologists who make every decision defensible. They're people who understand that working with AI isn't about replacement—it's about elevation.

The Future That's Already Here

We stand at an inflection point. The convergence of autonomous AI, regulatory frameworks, and quantum computing creates both unprecedented opportunity and existential risk. Companies that move now—thoughtfully, securely, systematically—will operate at a fundamentally different level than those that wait.

But "moving" doesn't mean reckless deployment. It means starting with one well-bounded agent. It means building monitoring before scaling. It means treating security as an enabler, not an afterthought. It means thinking in systems, not components.

Whether you're deploying your first agent or orchestrating hundreds, the principles remain constant. Remember the three stages of scaling from Chapter 9? Every organization moves from the Honeymoon stage through the Coordination Crisis to become an Autonomous Orchestra. But the companies that reach Orchestra status don't stop there—they become the ones setting industry standards and teaching others to scale safely.

Trust nothing. Verify everything. Build boundaries that enable. Make security invisible. Scale systematically.

Your Decision Point

Every business leader faces the same choice today that internet pioneers faced in 1995. You can see the transformation coming. You understand the potential. The question is: what will you do about it?

Some will wait for perfect clarity that never comes. They'll form committees, commission studies, and watch competitors pull ahead. Others will rush forward without foundations, deploying agents that become liabilities rather than assets. Both paths lead to obsolescence.

But there's a third path—the one this book illuminates. Build on Zero Trust foundations from day one. Start small but think big. Move fast but move securely.

The Call That Matters

I'll leave you with one final thought. The age of agentic AI isn't coming—it's here. In boardrooms and break rooms, in factories and hospitals, autonomous agents are making decisions that shape our world. The only question is whether those agents are secure, bounded, and aligned with human values.

You have everything you need to answer that question correctly. The Agentic Trust Framework provides your foundation. Industry patterns show what works. Implementation roadmaps guide your journey. Crisis playbooks prepare for the inevitable.

The companies that win won't be those with the most agents or the most sophisticated AI. They'll be those who master the balance between autonomy and control, between innovation and security, between machine capability and human wisdom.

Your autonomous agents are waiting. Build them on Zero Trust foundations. Scale them with MassiveScale principles. Deploy them with confidence. Transform your business while securing its future.

Welcome to the age of agentic AI. Build it right. Build it securely. Build it now.

The revolution isn't tomorrow—it's today. And you're ready to lead it.

Glossary of Terms

Here's a plain-English guide to the key terms everyone can use to understand AI agents:

A

Agentic AI
AI that doesn't just answer questions but actually gets things done—making decisions, taking actions, and completing work-flows without constantly asking for guidance.

Agentic Infrastructure
The technical foundation that supports your AI agents—servers, networks, and software platforms they need to operate reliably and securely.

Agentic Trust Framework
A specialized security model designed specifically for manag-ing AI agents, ensuring they remain trustworthy and aligned with your business goals.

AI Agents
Individual pieces of software that can think, decide, and act independently—like digital employees with specialized skills and responsibilities.

AI Security

Protecting your AI systems from hackers, malfunctions, and misuse, ensuring only authorized people can access them and sensitive data stays secure.

AI Transformation

The big-picture change when a company adopts AI across operations, rethinking how the entire business works with intelligent digital workers.

Artificial Intelligence (AI)

Technology that analyzes patterns in data to generate useful outputs—from answering questions to predicting trends—without being explicitly programmed for each specific task. While AI can simulate intelligent behavior, it works through sophisticated pattern matching rather than actual thinking or understanding.

Autonomous AI

AI that operates independently without constant human supervision, adapting to new situations and making decisions based on changing conditions.

Autonomous Digital Workers

AI employees who work independently, adapting to new situations and making judgment calls within their expertise without following rigid scripts.

B

Business Case

Your compelling financial argument for investing in AI, showing exactly how much money it will save or generate and how quickly you'll see returns.

Business Leaders

Executives and decision-makers who need to understand AI well enough to make smart strategic choices without becoming technical experts.

Business Value

The measurable benefits your company gets from AI—increased revenue, reduced costs, faster service, or better decision-making that you can track and prove.

C

CAIO (Chief AI Officer)

The executive responsible for your company's AI strategy and implementation, ensuring AI agents deliver value while working safely with the CISO to prevent security disasters.

Change Management

Helping your organization adapt to AI without chaos, planning for how people's jobs will change and supporting your team through the transition.

CISO (Chief Information Security Officer)

The executive responsible for protecting your company's digital assets, ensuring AI agents don't expose data or cause security breaches.

Competitive Advantage

The edge you gain over competitors by using AI more effectively—serving customers faster, predicting trends better, or operating more efficiently.

Compromised Agents

AI agents that have been hacked or corrupted, potentially stealing data, disrupting operations, or making harmful decisions against your interests.

Crisis Management

Your plan for handling AI emergencies—what to do when agents malfunction, get hacked, or start making decisions that could harm your business.

D

Data Movement

How information flows between your AI agents and business systems—customer data, inventory records, and other information agents need to work effectively.

Demand Spikes

Sudden increases in customer demand that AI agents can handle by scaling up instantly and processing thousands of requests simultaneously.

Digital Workers

AI-powered software that performs work tasks traditionally done by humans, adapting and learning rather than following rigid automation scripts.

F

Future-Proofing

Building AI systems that can adapt and grow as technology advances, designing flexible systems that won't become obsolete in two years.

H

Human Planners
Traditional employees who make strategic decisions and analyze data, whose roles evolve to focus on high-level strategy and managing AI systems.

I

Industry-Specific Playbooks
Customized AI implementation strategies for different business types, providing tested approaches and use cases for your particular industry.

Inventory Holding Costs
The expense of storing unsold products, which AI agents can reduce by predicting demand accurately and optimizing stock levels in real-time.

M

Misconfigured Permission
When AI agents get wrong access levels—either too much access (security risk) or too little access (can't do their jobs effectively).

P

Performance Metrics
Specific measurements tracking how well your AI agents perform—response times, accuracy rates, cost savings, or revenue generated.

Permission Management

Controlling what your AI agents can access in business systems, ensuring they can do their jobs without creating security vulnerabilities.

R

Risk Management

Identifying and preparing for potential AI problems—technical malfunctions, poor decisions, or security breaches—and using AI smartly and safely.

Rogue Agents

AI agents behaving in unintended ways that could harm your business, needing to be identified quickly and brought back under control.

S

Security Protocols

Rules and procedures keeping AI systems safe from threats, including technical measures like encryption and operational procedures like security audits.

Security Teams

IT professionals protecting your digital assets, who need to understand AI vulnerabilities and protect against AI-specific threats.

Speed of Light Operations

AI agents' ability to make decisions and take actions almost instantaneously, enabling new business models impossible with human-only operations.

Supply Chain
The network bringing products to customers, which AI agents can optimize by predicting demand, coordinating shipments, and adapting to disruptions.

System Access
How AI agents connect to business systems like databases and software, ensuring they get needed information while maintaining security.

Z

Zero Trust
Security philosophy assuming nothing should be automatically trusted, continuously verifying that every user, device, and AI agent is authorized.

Zero Trust AI Ecosystem
A comprehensive security environment designed for businesses using AI agents, combining zero trust principles with AI-specific protections.

Zero Trust Framework
A structured approach to implementing zero trust security, providing tested methods and guidance for building secure AI environments.

Sources

Comprehensive Bibliography

Market Research Reports

Compliance Automation Market

- *Future Market Insights. "Compliance Automation Tools Market Report 2024-2034." July 2024. Market analysis showing $1.45 million average savings from compliance automation adoption.*
- *Mordor Intelligence. "Compliance Management Software Market - Growth, Trends, and Forecasts (2024-2029)." 2024.*

Zero Trust Security Market

- *MarketsandMarkets. "Zero Trust Architecture Market - Global Forecast to 2028." Report No. TC 8499. November 2023. Projects market growth from $17.3 billion (2023) to $38.5 billion (2028).*
- *MarketsandMarkets. "Zero Trust Security Market Size, Share & Analysis Report to 2029." September 2024. Updated projections: $36.5 billion (2024) to $78.7 billion (2029).*
- *Grand View Research. "Zero Trust Security Market Size, Share & Trends Analysis Report 2021-2028." July 2021. CAGR analysis of 15.2%.*
- *Mordor Intelligence. "Zero Trust Security Market - Growth, Trends, and Forecast (2025-2030)." 2025. Market size analysis and regional breakdowns.*

- Roots Analysis. "Zero Trust Security Market Size & Share Report, 2024-2035." November 2024. Long-term market projections.

Industry Studies and Surveys

Compliance Statistics

- Hyperproof. "50+ Compliance Statistics to Inform Your 2024 Strategy." August 2024. Key finding: 18% automation rate for IT risk data collection.
- Drata. "115 Compliance Statistics You Need To Know in 2023." 2023. Comprehensive compliance metrics.
- Sprinto. "Top 120 Compliance Statistics (Updated 2025)." February 2025. Updated industry benchmarks.
- Secureframe. "110 Compliance Statistics to Know for 2025." October 2024. Survey data on compliance automation adoption.
- NAVEX. "2024 State of Risk & Compliance Report." 2024. Enterprise compliance trends and challenges.
- Thomson Reuters. "Cost of Compliance 2022 Report." 2022. Global compliance cost analysis.
- Capgemini Research Institute. "Rise of agentic AI: How trust is the key to human-AI collaboration." 2025. Projects $450 billion economic value by 2028; findings on trust and deployment challenges.

Cybersecurity Research

- IBM Security. "Cost of a Data Breach Report 2024." July 2024. Key findings: 40% of breaches involve multi-environment data.
- IBM X-Force. "Threat Intelligence Report 2024." February 2024. Annual threat landscape analysis.

- Chainalysis. "Crypto Crime Report 2025." December 2024. Analysis of $2.3 billion in crypto-related cybercrime.
- Ponemon Institute and Globalscape. "Compliance Cost Reduction Best Practices Study." 2023. Twelve practices for reducing compliance costs.
- Cyber Security Tribe. "Experts Reveal How Agentic AI Is Shaping Cybersecurity in 2025." May 17, 2025. Key finding: 59% of CISOs actively implementing agentic AI.

Organizational Transformation

- McKinsey & Company. "Common pitfalls in transformations: A conversation with Jon Garcia." McKinsey Insights, March 2022.
- Evergreen, Brian. "Autonomous Transformation: Creating a More Human Future in the Era of Artificial Intelligence." Wiley, 2023. ISBN 978-1119985297. Chapter on Problem Solving versus Future Solving framework.
- Bornet, Pascal. "Agentic Artificial Intelligence: Harnessing AI Agents to Reinvent Business, Work and Life." Irreplaceable Publishing, 2025. ISBN 979-8992833645. Implementation success factors and challenges from analysis of hundreds of agentic AI deployments.

Academic Research

- Harvard Kennedy School and Avant Research Group. "Evaluating Large Language Models' Capability to Launch Fully Automated Spear Phishing Campaigns: Validated on Human Subjects." arXiv:2412.00586. December 2024. Key finding: AI phishing achieves 54% click-through rate.
- Harvard Kennedy School Misinformation Review. "How Spammers and Scammers Leverage AI-Generated Images on Facebook for Audience Growth." February 2025.

- *Cyentia Institute and SecurityScorecard. "Third-Party Risk Management Research Report." 2023. Analysis of supply chain vulnerabilities.*

Government and Regulatory Sources

United States

- *Cybersecurity and Infrastructure Security Agency (CISA). "Emergency Directive 24-01: Mitigate Ivanti Connect Secure and Ivanti Policy Secure Vulnerabilities." January 2024.*
- *Department of Health and Human Services. "HIPAA Enforcement Statistics 2024." 2024.*
- *Federal Trade Commission. "Consumer Sentinel Network Data Book 2024." February 2024.*

International

- *Financial Services Agency of Japan. "Alert on Unauthorized Securities Trading." April 2025. Report on $350 million in unauthorized trades via compromised systems.*
- *European Union Agency for Cybersecurity (ENISA). "Threat Landscape 2024." October 2024.*

Technology Vendor Reports

- *Accenture. "2022 Compliance Risk Study." 2022. Finding: 95% of businesses building compliance culture.*
- *Zscaler. "State of Zero Trust Transformation 2023." 2023. Cloud security and Zero Trust adoption trends.*
- *UserCentrics. "Compliance Automation: What You Need to Know and How to Get Started." April 2025. ROI analysis of compliance automation.*
- *CrowdStrike. "Global Threat Report 2024." February 2024. Advanced persistent threat analysis.*

- *Fortinet. "Compliance Automation Guide." 2024. Technical implementation guidance.*
- *Microsoft. "Microsoft Build 2025: The age of AI agents and building the open agentic web." The Official Microsoft Blog, May 19, 2025.*
- *Cisco. "Cisco Transforms Security for the Agentic AI Era, Further Fusing Security into the Network." Cisco Newsroom, June 10, 2025.*
- *Google Cloud. "The dawn of agentic AI in security operations at RSAC 2025." Google Cloud Blog, April 28, 2025.*
- *CyberArk. "The Agentic AI Revolution: 5 Unexpected Security Challenges." CyberArk Blog, February 28, 2025.*
- *Beyond Identity. "The Hidden Cost of Enterprise AI: Identity-First Security for LLMs and Agents." 2025. Analysis of AI identity risks and security architecture.*

News and Media Coverage

Data Breach Incidents

- *Anthropic. (2025, January 30). Project Vend: Can Claude run a small shop? (And why does that matter?).*
- *Kerravala, Z. (2025, June 5). Surfing the AI wave with Zero Trust everywhere: Five takeaways from CEO Jay Chaudhry's keynote at Zscaler's Zenith Live. SiliconANGLE.*
- *Reuters. "OpenAI's internal AI details stolen in 2023 breach, NYT reports." July 5, 2024.*
- *TechCrunch. "Evolve Bank says ransomware gang stole personal data on millions of customers." July 9, 2024.*
- *SecurityWeek. "Evolve Bank Data Breach Impacts 7.6 Million People." July 9, 2024.*
- *BBC News. "Hong Kong: Finance worker pays out $25m after video call with deepfake CFO." February 2024.*

- CNBC. *"Hackers stole twice as much crypto in the first half of 2024."* July 9, 2024.
- The Record: Recorded Future News. *"Japan warns of hundreds of millions of dollars in unauthorized trades from hacked accounts"* April 21, 2025
- SC Media. *"Your AI agent is an overprivileged intern with no HR file."* 2025. Critical analysis of agentic AI security challenges and vendor responses.
- Fowler, Geoffrey A. *"Is OpenAI's Operator, a new AI 'agent,' ready to help in the real world?"* The Washington Post, February 7, 2025. Technology section.

Market Analysis

- CSO Online. *"AI gives superpowers to BEC attackers."* 2024. Analysis of AI-enhanced business email compromise.
- Malwarebytes Blog. *"AI-supported spear phishing fools more than 50% of targets."* January 7, 2025.
- SiliconANGLE. *"AI agent-powered compliance automation startup Norm Ai raises $48M."* March 11, 2025.
- Axios. *"Security teams start to embrace agentic AI."* March 27, 2025. Analysis of CrowdStrike and Trend Micro agentic AI deployments.
- CIO. *"88% of AI pilots fail to reach production—but that's not all on IT."* October 2024.
- BreakingDefense. *"Anthropic, Google and xAI win $200M each from Pentagon AI chief for 'agentic AI'."* July 2025. Pentagon investment in commercial agentic AI platforms.
- PYMNTS. *"Google Adds Agentic AI Capabilities to Timesketch Cybersecurity Platform."* July 15, 2025. Analysis of Google's AI-powered cybersecurity advances.
- AWS Newsroom. *"AWS Invests Additional US$100m in Generative AI Innovation Center for Agentic AI*

Development." 2025. Announcement of expanded funding for enterprise agentic AI deployment.

- SiliconANGLE. "CrowdStrike's Daniel Bernard on AI-Powered Cybersecurity Innovation." Interview at AWS Summit NYC. 2025. Discussion of agentic AI in cybersecurity defense.
- KPMG. "AI 2025 Quarterly Pulse Survey." 2025. Finding: 65% of companies piloting agentic AI.
- Forbes. "Why Absorptive Capacity Is The Hidden Key To Thriving With Agentic AI." 2025. Analysis of organizational readiness factors.
- "SoftBank Plans AI Workforce of 1 Billion Agents This Year." 2025. CEO announcement of massive internal agent deployment and elimination of human programming.
- GitGuardian. "RSA Conference 2025: How Agentic AI Is Redefining Trust, Identity, and Access at Scale." 2025. Analysis of AI security discussions featuring Anthropic, OpenAI, Meta, and Zscaler perspectives.
- StateScoop. "Oklahoma's cybersecurity chief embraces 'scary' AI agents to fight AI-powered attacks." 2025. Profile of state government deployment of autonomous security AI.

Professional Organizations and Industry Bodies

- American Medical Association. "Physician Practice Benchmark Survey 2023." 2023. Administrative burden statistics.
- National Practitioner Data Bank. "2023 Annual Report." 2023. Medical malpractice statistics.
- Financial Industry Regulatory Authority (FINRA). "Regulatory Notices on Unauthorized Trading." Various dates.

- *International Organization for Standardization (ISO). "ISO/IEC 42005:2025 - AI System Impact Assessment." 2025.*
- *Open Web Application Security Project (OWASP). "OWASP Top 10 for LLM is now the GenAI Security Project and promoted to OWASP Flagship status." March 27, 2025.*
- *Open Web Application Security Project (OWASP). "Agentic AI - Threats and Mitigations." April 28, 2025.*
- *Prosci. "ADKAR: A Model for Change in Business, Government and our Community." Prosci Inc., 2025. Change management framework.*

AI Security Standards and Frameworks:

- *MITRE Corporation. "MITRE ATLAS (Adversarial Threat Landscape for Artificial-Intelligence Systems)." 2025. Available at: https://atlas.mitre.org.*
- *Coalition for Secure AI (CoSAI). "Open Source AI Security Tools and Frameworks." OASIS Open Project. 2025.*
- *National Institute of Standards and Technology (NIST). "Artificial Intelligence Risk Management Framework (AI RMF 1.0)." NIST AI 100-1. January 2023.*
- *International Organization for Standardization. "ISO/IEC 42001:2023 - Information technology — Artificial intelligence — Management system." December 2023.*
- *Cloud Security Alliance. "MAESTRO: AI Model Security, Evaluation, & Standardized Testing for Responsible Operations." 2024.*
- *Open Web Application Security Project (OWASP). "Artificial Intelligence Security Verification Standard (AISVS)." Version 1.0. 2024.*
- *AI Incident Database. "AIAAIC Repository: AI, Algorithmic, and Automation Incidents and Controversies." Partnership on AI. 2025.*

Specialized Security Research

- Cyvers. *"Annual Crypto Security Report 2024."* December 2024. Analysis of $2.3 billion in crypto losses.
- TRM Labs. *"Crypto Crime and Compliance Report H1 2024."* July 2024. Cryptocurrency security trends.
- Verizon. *"2024 Payment Security Report."* 2024. PCI compliance statistics.
- Sophos. *"State of Ransomware in Financial Services 2022."* 2022. Sector-specific threat analysis.
- Lasso Security. *"Top 10 Agentic AI Security Threats in 2025 & Fixes."* July 2025. Analysis of memory poisoning and tool misuse as primary agentic AI threats.

Consulting and Advisory Firms

- Gartner. *"Top 5 Priorities for Compliance in 2025."* December 2024. Enterprise compliance trends.
- Forrester Research. *"The State of AI Security 2024."* 2024. AI implementation security analysis.
- Deloitte. *"Global Risk Management Survey, 11th Edition."* 2019. Enterprise risk management benchmarks.

Note on Sources: All sources were verified as of August 2025. Given the rapidly evolving nature of AI and cybersecurity, readers are encouraged to check for updated versions of annual reports and market studies. URLs have been omitted from this print bibliography but are available at [book companion website].

Accessibility: For digital copies of sources, updated statistics, and direct links to reports, please visit [book website/resources]. QR codes for specific citations are available in the enhanced digital edition.

www.ingramcontent.com/pod-product-compliance
Lightning Source LLC
Chambersburg PA
CBHW040752220326
41597CB00029BA/4731